Leabharlanna Chontae Lú

3 0009 00582503 4

IVORY, APES AND PEACOCKS

Alan Root was born in London in 1937 but moved to Kenya as a young boy. He dropped out of school at sixteen but soon found himself behind the camera. He married Joan Thorpe in 1961 and together they produced an array of groundbreaking wildlife films including *Baobab: Portrait of a Tree*, commissioned by David Attenborough, *Safari by Balloon* and *Mysterious Castles of Clay*, which was nominated for an Oscar.

Alan has won over sixty awards during his career, including three Lifetime Achievement Awards and an OBE. He flies aeroplanes, balloons and a helicopter that he uses for various conservation projects, and now lives on the Lewa Wildlife Conservancy in northern Kenya with his wife and two young sons.

D1351395

ALAN ROOT

Ivory, Apes and Peacocks

Animals, Adventure and Discovery in the Wild Places of Africa

VINTAGE BOOKS
London

COUNTY LOUTH
LIBRARY SERVICE
Acc No 5825034
Class No 920 Roo
Invoice No 400152313
Catalogued 15/11/13
Vendor Omanolys

Published by Vintage 2013

2 4 6 8 10 9 7 5 3 1

Copyright © Alan Root 2012

Alan Root has asserted his right under the Copyright, Designs
and Patents Act 1988 to be identified as the author of this work

This book is a work of non-fiction based on the life, experiences and
recollections of the author. In some limited cases names of people, dates,
sequences or the detail of events have been changed to protect the privacy
of others. The author has stated to the publishers that, except in such
minor respects not affecting the substantial accuracy of the work,
the contents of this book are true

This book is sold subject to the condition that it shall not,
by way of trade or otherwise, be lent, resold, hired out,
or otherwise circulated without the publisher's prior
consent in any form of binding or cover other than that
in which it is published and without a similar condition,
including this condition, being imposed on the subsequent purchaser

First published in Great Britain in 2012 by
Chatto & Windus

Vintage
Random House, 20 Vauxhall Bridge Road,
London SW1V 2SA

www.vintage-books.co.uk

Addresses for companies within The Random House Group Limited
can be found at: www.randomhouse.co.uk/offices.htm

The Random House Group Limited Reg. No. 954009

A CIP catalogue record for this book
is available from the British Library

ISBN 9780099555889

The Random House Group Limited supports the Forest Stewardship
Council® (FSC®), the leading international forest-certification organisation.
Our books carrying the FSC label are printed on FSC®-certified paper. FSC
is the only forest-certification scheme supported by the leading environmental
organisations, including Greenpeace. Our paper procurement policy
can be found at www.randomhouse.co.uk/environment

Typeset in Melior by Palimpsest Book Production Limited, Falkirk, Stirlingshire
Printed and bound by CPI Group (UK) Ltd, Croydon CR0 4YY

For Fran, who I dream with, laugh with and love.
For Myles and Rory, who keep me young, may there always
be wild creatures and places in which to find joy and solace.
And for everyone who strives to save the natural world that
I have been so privileged to record and enjoy.

Quinquireme of Nineveh from distant Ophir,
Rowing home to haven in sunny Palestine,
With a cargo of ivory,
And apes and peacocks,
Sandalwood, cedarwood, and sweet white wine.

'Cargoes', John Masefield

Contents

Sketches by Wolfgang Weber

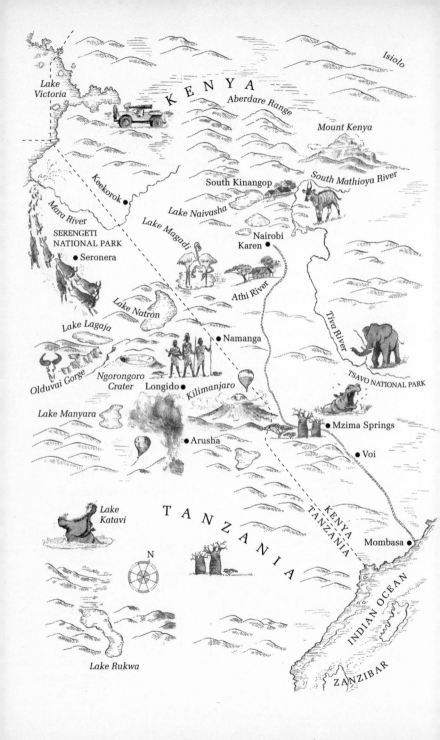

Illustrations

Sketches by Wolfgang Weber

IVORY, APES AND PEACOCKS

Prologue

'*Muzungu!* White man! Why are you lying down?' My interrogator was an elderly African who fortunately spoke not his tribal dialect, but KiSwahili, which I could understand. Not very well, however, as he'd had more than a skinful of the local palm wine and could barely stand, let alone speak. I was lying down because I was pinned under a motorbike, but it was evidently a stretch for him to understand my predicament. For some fifteen minutes I had been struggling in the slippery mud to lift the bike off my leg, but I just didn't have the strength. I was exhausted. Two days earlier I had left my camp deep in Africa's biggest and wildest rainforest, the Maiko forest in Zaire, where, with the help of a group of Mbuti pygmies, I had spent two weeks in search of one of the world's rarest birds – the Congo peacock. I had walked a full day to the straggling group of a dozen huts called Balobé, where, after a sleepless night on a pygmy-sized wooden bed in a hut full of rutting goats, I had collected the trail bike that I had left there two weeks earlier and set off on a narrow footpath.

I had been riding this treacherous trail for almost eight hours when I hit a stretch of particularly deep mud and fell, with the bike and its heavy load landing on top of me. I was completely done in and felt as if I were pinned under a Harley.

1

I had passed a man about a mile back, the only human I had seen since leaving Balobé. He hadn't returned my greeting but he had been headed this way and was my only hope of getting help. But this was Zaire, a country with a history of ruthless exploitation by white men and equally envenomed retribution. It was not beyond the bounds of possibility that, in my helpless state, I would be relieved of my watch, clothes, cameras, bike and perhaps even my life. No one would see, no one would know.

It was a biker's nightmare. Over the years, trees of all sizes had crashed down across the trail. The larger ones had to be climbed over, or crawled under, with the bike unloaded, then lifted or dragged behind. The smaller fallen trees had been cut to clear the path, leaving a gap just wide enough for passage on foot. Hidden by vines and vegetation, the sharp ends were perfectly placed to catch my ankles and knees, and soon the cuts and bruises I had gathered on the inward journey to the camp were reopened and enlarged. At every river bridged by a fallen tree I'd had to unload and carry my cameras and gear over, then manoeuvre the bike across the narrow slippery trunk, very aware that this was no place to be alone.

'Why? Why are you lying in the mud?'

Fortunately my new friend was a happy drunk, if also rapturously beyond reason. He leaned over me, swaying perilously and, coughing wheezily, he bathed me in alcohol fumes so dense I was relieved the engine wasn't running.

'*Kwa nini?* Why?' Oh man, this was *not* the local Automobile Association.

Detecting a certain attention deficit, I carefully explained that I was not lying there by choice and that I would like some help, mentioning a sum that would keep him in palm wine for a month. Unfortunately he had passed the point where money was of any interest. It was getting late and he had an agenda, no matter how hazy.

'Rain is coming,' he offered helpfully as he started to move off. Then, over his shoulder by way of encouragement, 'Soon.'

'Please, please, old man, come back, I need your help! I'm not lying here because I'm sleepy – I can't get up by myself.'

Being unable to get up seemed to strike a chord with him – something he had experienced personally and might again at any minute. So he shuffled slowly back and finally, down on his knees and with great difficulty, he was able to take the weight off my leg long enough for me to drag it clear of the bike and stand up. I thanked him profusely and gave him a three-inch wad of grubby Zairean notes that represented a couple of dollars along with a handful of aspirins for what he described as a 'noise like bees' in his head – if that was all, I thought, he was getting off lightly. As the first fat raindrops fell, he staggered off into the endless gloom of the forest and I wondered how far he had to go, and if he was even heading in the right direction. I bent down to lift the bike, slipped, and fell again. I was exhausted and lay there clawing fiercely at the mud thinking, I'm over sixty – what the hell am I doing here?

In 1909, just ten years after Joseph Conrad had so vividly portrayed the treasures and torments of the Belgian Congo in his haunting *Heart of Darkness*, two men made the same slow trip up the Congo river to Stanleyville, the nightmarish 'Central Station' of Conrad's novella. From there they travelled east with two hundred porters for thirty days and established a base at Medge, home to the Mangbetu tribe, who were known for tightly binding the back of their skulls, giving them greatly elongated heads. Unlike so many men who travelled this way, they were not colonial officials, nor ivory-hunting freebooters: they were naturalists, intent upon exploring the least-known and most biologically diverse forests on the planet. Financed by the American Museum of Natural History in New York, the expedition was led by Herbert Lang, a talented German-born taxidermist described as 'a man of almost superhuman energy'. His only assistant, nineteen-year-old James Chapin, would later become the finest ornithologist ever to work in Africa. Despite the hardships of rainforest life, theirs was to prove a tremendously rewarding exploration. For the next five years, with few means of communication – unaware that in the even darker heart of Europe the world was moving towards war – they collected literally tons of invaluable zoological and anthropological specimens, many of them new to science.

Somewhere in those dark forests, in that ocean of every shade of green, hid the okapi, an animal that was top of Lang's list. In 1890 Henry Stanley, the celebrated explorer, had first reported a horse-like animal living in the forest. Then in 1901 Sir Harry Johnston, the High Commissioner of Uganda, had sent an unidentified piece of striped skin to the British Museum, which prematurely declared it to be a new species of zebra. Johnston set out, with ambitions to have a new species named after him, to the Ituri forest in the Belgian Congo to track down the 'zebra'. There he became thoroughly confused when his Mbuti guides showed him the hoof prints of the animal he was seeking. They were cloven, so it could not be a horse of any kind. Johnston continued his search, and when he finally collected some skin and a skeleton the mystery was resolved. The animal he had discovered was indeed the size of a horse, but there the similarity ended. This creature had an elongated neck to allow it to reach for leaves, but not so long as to hinder its progress through the forest. It was rich chestnut in colour, with narrow, chalk-white horizontal stripes across its rear end, stumpy, hair-covered horns and a tongue long enough to lick its eyes – it was clearly a member of the giraffe family. The local native name was *o'api*, and Johnston's efforts were rewarded when the animal was scientifically named, as he had hoped, *Okapia johnstoni*. A few years later

Lang's great ambition was to collect complete specimens suitable for a museum exhibit.

Along with the okapi, somewhere deep in the forest, creeping along the dense margins of whisky-coloured, tannin-stained streams, lived a small, secretive predator that Lang's expedition would also be the first to collect. A cat-sized, low-slung, mongoose-like carnivore, with dense orange-red fur, black legs and tail, and white markings round its eyes, it was an enigma that still stokes taxonomical dispute to this day. A hunter brought in the dead creature, but nothing else was known about it and even today it is represented in the world's museums by only a handful of skins from animals killed in hunters' traps or by their dogs. No naturalist had ever observed one alive and all that was known was that it had partially webbed feet and the remains of fish were found in its stomach, so it was named the aquatic genet. Lang's diligent assistant James Chapin was presented with many such mysteries during the five-year expedition and one of them was particularly puzzling. He had collected hundreds of bird species, but there was one single feather that he had come across in a native's headdress that he couldn't identify. A glowing rufous orange with black barring, it seemed to belong to a game bird such as a guinea fowl or francolin, but it was too big, almost pheasant-like, yet Chapin knew that no true representatives of the pheasant family occurred in Africa. After identifying and cataloguing hundreds of specimens – indeed, every feather – in his vast collection, Chapin was left with that one mysterious quill. It tormented him and, with his impeccable museum conscience, he kept it in a safe place.

Twenty-three years passed and James Chapin found himself in Belgium, researching in the Congo Museum at Tervuren, near Brussels. Passing along a seldom-used corridor one day he spotted, stowed away on top of a cupboard, two badly stuffed and dusty birds labelled simply 'Pavo cristatus, immature'

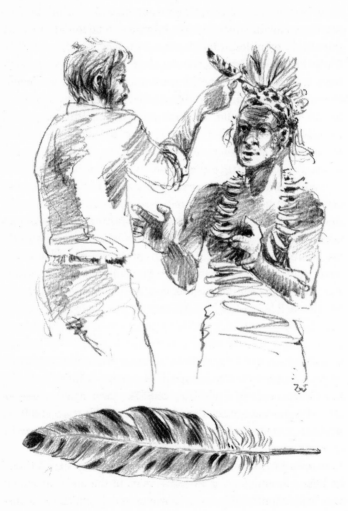

– i.e. young common peacocks. Blowing off the dust, Chapin immediately recognised the source of his mysterious feather and, when the quills proved a perfect match, he had a dramatic discovery on his hands, the first true pheasant – with the characteristic metallic lustre on its plumage – to be found in Africa, thousands of miles from the large pheasant family in Asia. Those two dusty birds had obviously been mislabelled by some long-dead curator, so checking the records to find out where they had come from, Chapin returned to the Congo in 1937, twenty-eight years after his first expedition with Lang, to try to collect specimens of this new species. But while his hunters managed to bag several of what was now called the Congo peacock, Chapin himself never saw one in the wild.

As another world war came and went, the Congo peacock disappeared back into the vast, dark forests of the Congo. In the late 1940s a renowned animal collector named Charles Cordier managed to obtain several live specimens. His birds, too, were caught by his African trappers so, like Chapin, he never saw a peacock in the wild. Fast forward to 1978, when *Audubon* – an ornithological magazine – sponsored an expedition to search for the bird. Again it had no success.

Some years after that it was my turn. I flew into a remote forest airstrip for a week-long search with three of Africa's top bird men – Ian Sinclair, Don Turner and Don's son Steve. Although they added many species to their legendary life lists, the Congo peacock eluded us, too. The bird had become an almost mythical creature, the ornithologist's holy grail, and there was still no record of a naturalist ever sighting one in the wild.

I grew up in Africa and by the time of that 1986 expedition, I had been filming for almost thirty years in the open savannahs and bush country of Kenya, Tanzania and Uganda. The dense rainforests of the Congo, filled with fabled creatures, presented a new and exciting challenge. As a boy, I had been entranced by my otherwise rather boring English teacher reading John Masefield's poem 'Cargoes' and had often imagined myself sailing with that romantic ship, the quinquireme of Nineveh, with its rows of muscular oarsmen and exotic cargo of ivory, apes and peacocks. In later life I had spent a lot of time with elephants and the ivory that is their glory and curse, and had lived with Africa's greatest apes, the mountain gorillas. Now I wanted to complete that trio of boyhood dreams and add the peacock – not the familiar, fantailed denizen of India and hotel gardens – but Africa's own, deeply mysterious bird, the Congo peacock. Inspired by the expeditions of these early naturalists – Johnston, Lang and Chapin – I was determined to find and film the okapi, the aquatic genet, the peacock, and the many other strange and rare forest creatures I had read about. But when I made a reconnaissance flight over the vast Congo forests, it instantly became obvious why so many species had eluded discovery for so long. I flew for hour after hour over seemingly endless and featureless dense vegetation. For easier handling when flying my own small plane on these difficult flights, I had cut my unwieldy flying maps down into A3-sized pages, which I kept in a folder with each page labelled. One such page, covering over forty thousand square miles, has only three features on it that could charitably be called towns and is simply labelled 'The Big Broccoli'.

After that initial recce to Zaire, as the Congo was now called, I knew I had to return. I had no idea how, though, especially given the size of those forests and the continued political chaos in that country. In 1986 I eventually secured a contract with *Survival*, a wildlife television series in the UK, and *National Geographic* in the States, to make four one-hour films in Zaire, which would give me at least eight years in this glorious but godforsaken country: eight years in which to seek out the elusive creatures I had set my heart on filming. That's how I had got to be lying there in the mud. Getting there had been a long, long journey.

8

1

Into Africa

George VI, King of England, Emperor of India, Defender of the Faith etc. etc., was crowned on the same day I was born, 12 May 1937. In a publicity-grabbing display of tradition and power, the parades and ceremonies held that day tended to draw attention away from me and my mum. However, my mother's patriotic efforts in giving birth that day did not go unnoticed. My arrival, along with quite a few other Coronation Babies, was greeted in the press, and the Lord Mayor presented Mum with five pounds and a royal-blue pram. She was always going to call me Alan, but in a fever of royalism and gratitude she tacked on the King's moniker, George Windsor, mercifully leaving out the Albert Frederick Arthur bit. After that, his life and mine soon drifted apart, but decades later I was destined to meet his daughter, by then the Queen and, totally bereft of small talk, I told her the story. I think she was amused.

Born in London's East End, within the sound of the Bow Bells, I was a cockney. My parents, Violet and Edwin, were working-class folk from London, both from large families. My dad was a manager at Peck's, a factory that produced fish paste and sundry other pungent products, which featured large on menus at home (fortunately one of my many aunts balanced our diets by working in a chocolate factory). Two years after my birth we were joined by my sister Jacqueline and shortly

after that by Hitler. I was too young to realise the effect that the declaration of war must have had on my parents, with their newly minted family, until almost sixty years later. Then, as the elderly father of two small boys, my first children, I sat crying as I watched the Twin Towers come down and knew that the world had changed for ever. My parents' experience must have been far more immediate and much worse.

My only memories of life in London are of the sound of sirens and bombs, of a night we spent sheltering in the Underground and my mother's exaggerated cheerfulness. Our street was bombed, as was the church of St Mary-le-Bow, its bells silenced until 1961, when they heralded a new generation of cockneys. Along with many other women and children, we were evacuated from the danger and climbed down from the train at the chilly coastal town of Eastbourne, Mum holding a cardboard suitcase in one hand and my sister in the other, and me with a canvas holdall. Nothing had been organised, and there was no one to meet us – all we knew was that we had to find ourselves somewhere to stay. It was drizzling, with an icy wind coming off the Channel, as we trudged the wet streets trying to look hopeful. Then came a soft voice: 'Come on, love, you come and live with me. I've got some kids that age' and with no more than that we were taken in by Elsie Hoare, a warm and wonderful woman who shared her home and family with us for the next year.

A walk on the Eastbourne beach, with its frozen pebbles and icy sea, can be a pretty grim experience. But there were shells and crabs and pieces of smoothed and rounded green or red glass; emeralds and rubies dropped by some passing pirate – for a kid who had only known a broken city it was paradise.

In a way we were blessed. My father had serious varicose veins, swollen ropes that wrapped his legs like a strangler fig on a forest tree and had to be tightly bandaged several times a day. As a result he could not be sent off to fight, so he stayed in London, assigned to fire-watch duty on the roof of Peck's factory, and in the long, lonely hours between bombing raids, he carved wooden toys for me: wartime toys; wonderfully realistic tanks and trucks, planes and V1 rockets, which he painted with camouflage colours and the appropriate insignia; little houses with a garage, a lawn and a tree, the kind of house that I am sure he dreamed of living in one day when it was all over. I have most of them still and my boys much prefer them to modern plastic stuff.

Suddenly news came that my father had found a house for us in Marlborough, a pretty little village surrounded by rolling downs, where a great white horse carved into a chalk hillside had badgers denning under its belly. For a child it was a magical place of peewits and larks, in a wide sky unsullied by barrage balloons and the trails of warring aircraft, a place of dark bluebell woods and the silvery River Kennet from which fat fish rose to the grasshoppers I threw them. School was a couple of miles away, with a walk through a graveyard that in the dark of winter was a heart-pounding start and end to the day.

My dad came down from London to visit us when he could and brought me a pair of heavy binoculars he could spare from his fire-watch work, and I spent all my free time birding and egg collecting. With today's population of small boys almost outnumbering that of hedge-nesting birds, egg collecting has rightly been banned, but back then it seemed a relatively harmless pursuit that introduced many great naturalists to their passion. I had a little book, then, and I have it still: *A Bird Book for the Pocket* by Edmund Sandars, and some of those illustrations are burned into my memory. That little book set my course and, after a lifetime spent with images of nature, those pictures – of the goldfinch, the long-tailed tit and the eggs of the yellow hammer – never fail to raise the hairs on the back of my neck as they did then.

On his visits from London my dad often brought chocolate from Aunt Cissy and always more of the dreaded 'Peck's Finest Fish Paste'. I have since learned that with its powerful aroma and taste it makes excellent bait for trapping small scavengers, but in those days I had to eat it. All we heard about the war was what Dad told us on these visits, of the new, silent 'buzz bombs' and the burning city. And then, overnight, the war came to Marlborough in the most surprising way: American soldiers. Suddenly we kids were chewing gum and lining up next to the huge Greyhound bus that had been converted into a mobile kitchen serving coffee and doughnuts. Doughnuts! After years of wartime food and fish paste, doughnuts were an unbelievable treat. Within a few months my egg collection was going well, and I was lucky enough to be invited to see the collection of a kindly old gentleman who lived in the village and had been collecting scientifically for decades. I don't recall his name, but I clearly remember what I learned that day. He was a giant of a man who had spent most of his life down a coal mine, hands like shovels, their creased skin indelibly marked by his trade – hands that I felt could crush the hardest of conkers, yet here he was, proudly, modestly and patiently telling me the story behind his greatest prize, the pea-sized egg of a goldcrest, England's smallest bird. That minute, fragile egg, nestled tenderly between the tips of those massive fingers, is a memory I hold dear to this day. Somehow, although I was only nine years old, I read him as a metaphor for the combination of strength and gentleness required to make a good man.

When the war was over, my father had had enough of England. He wanted something better for all of us and he had been working hard towards his escape. Dad was in the food production business, which was being given high priority by the government, and there were over eleven million square miles of Empire out there – India, Africa, Canada, Australia – all ready to welcome anyone with useful skills prepared to adopt a new, rough-and-ready country. My mother had been holding back the news until she was certain, then it exploded like one of those buzz bombs. Dad had got a job in Africa! He was sailing in a week and we would all follow a month later. I burst into tears at the news, sobbing that I had just learned *every one* of the British birds and now I was going to a country where that knowledge would be useless.

My dad was to be factory manager for a new Liebig's meat-packing plant in Kenya, twenty-five miles from Nairobi on the Athi-Kapiti plains, which in the forties were much like the Serengeti is today. In his first letters he wrote excitedly of the great herds of zebra and hartebeest migrating past his new house, and of lions attacking the cattle that were brought to slaughter every day. My father had little interest in birds, but like anyone first arriving in Kenya he was knocked out by the colours, variety and tameness of those in his garden. He wrote of firefinches, bee-eaters and sunbirds, and although I had no idea what any of them looked like, the names alone were bewitching and I began to think this might be a good move after all. He also wrote that on a clear morning he could see the glaciers of Mount Kenya from the front of the house, and the snows of Kilimanjaro from the back. (It was absolutely true, but Kenya's air can be crystalline and he failed to mention that they were both about ninety miles away.) That clinched it for me and when I packed my stuff for the trip I also took along my toboggan.

To get to Kenya back then, if you were in a hurry you sailed via the Mediterranean and the Suez Canal, which took just ten days. If such dizzying speed was not important you went round the Cape, which took a leisurely three weeks. We chose the latter and my first contact with African wildlife came when we bought a small tortoise in Durban that promptly laid nine eggs in the bunk-side drawer in my cabin. A sunburned grandmother on board gave us KiSwahili lessons every day and by the time we landed we had a few useful phrases in addition to those in our *Swahili Made Easy* phrase book – the most practical of which was 'Make sure you put plenty of salt on the lion's skin or the hair will fall out'. The book didn't say whether the treatment was for a flayed dead lion or a live one with alopecia. Whichever, half a century later I have yet to need it.

We docked at the old Arab seaport of Mombasa, and from there the train trip to Nairobi was all adventure. My sister and I investigated our sleeper compartment panelled with well-polished wood and gleaming brass, then a man wearing a red fez and long white nightshirt called a *kanzu* came walking down the corridor banging a gong and calling us to dinner. In the art-deco dining car we sat at tables set with silver and heavy white linen, and were served an even heavier three-course meal starting with that sustainer of Empire, Brown Windsor soup.

After climbing up through the lush coastal belt, with coconut palms, naked children and bare-breasted women along the track, darkness fell, and during the night we crossed the Tsavo river where in 1898 man-eating lions had delayed construction of the railway bridge for almost a year by killing and eating twenty-eight of the Indian labourers imported for the task. Next morning I was up in the dark, my head out of the open window, smoke smuts speckling my face, waiting for the dawn. The sun rose blood-red over the Athi plains, with a horizon as flat and distant as it had been at sea, giving me my first glimpse of the Africa my father had described. I shouted in amazement as herds of zebra and antelope raced alongside the train, and placid groups of giraffe looked down on us as we passed. (In days not too long gone, rich hunters had sat on the engine's cowcatcher, and the train would stop and wait while they went off and potted a lion or rhino. Despite these depredations both species were still common, but we failed to see any that morning.)

We didn't travel as far as Nairobi, but stopped at Athi River station, just a couple of miles from our new home. There was no platform, we simply climbed down next to the neat station-master's house with its immaculate little garden. Across a dusty road there was a single shop, with a couple of Maasai warriors standing outside complete with spears and long scabbarded knives, and then there was Dad with a Chevrolet station wagon, big enough for all our trunks and my toboggan. The drive to our new house took just ten minutes. We passed giraffe near the station, then enough gazelle, zebra, wildebeest and birds to make me reel.

The house was a modest tin-roofed colonial-style bungalow, with a veranda enclosed with mosquito mesh and a small arid lawn in front. Out back was a larger garden and a great variety of fruit trees – mangos, figs, pawpaws and guavas – all fruit I'd never seen before, let alone eaten. Running around exploring, I discovered our 'refrigerator' nestled outside under a shady mango tree: an ingenious double-walled cage of small-mesh wire with charcoal packed between the walls. A pipe dripped water down the walls and evaporation cooled the contents. And what contents! Jugs of milk, butter, pots of cream and trays of eggs, all items we had been deprived of while growing up in the shadow of rationing and the war.

My dad was too preoccupied by now to notice all this new variety and splendour around him – he was busy. Post-war England, broke and hungry, was desperate for meat and had instituted a government scheme called 'Bully for Britain'. My father's job was to produce as many cans of bully (corned beef) as possible and he went at it full bore. His first priority was to show me 'his' factory. Only I was invited – this was obviously male territory in his eyes. So the next morning, wrapped up against the pre-dawn cold, I walked the mile to the factory and found the pens from which the cattle were funnelled up to the killing floor. Here, two hundred animals a day were slaughtered three hundred and fifty days of the year. It was a job with little room for sentiment, but my dad was always concerned that no animals should suffer any avoidable pain or distress. Once the animals had been electrocuted, the carcasses were hung on hooks, then swiftly bled, gutted, skinned and trundled off to the various departments. The men here worked with blood halfway up their wellies as it poured down a huge drain to end up as blood meal. I was just ten, and remember being sickened by the sight and smell of so much blood; I wasn't convinced by Dad's assurance that the animals felt no pain, but I was certainly amazed by the alchemy going on in that huge building – the rapid conversion of meat on the hoof to meat in a tin, a jar or a bag. Once I had satisfied my father's apparent wish to see that I was man enough to survive the killing floor I never went back to the factory again.

Our house was the last in a row of six along a dusty road that led from the factory, and our new neighbours were a young South African couple with two small boys a bit younger than my sister and me. Soon after our arrival I was invited to go out with them for what they called a game drive. The boys and I piled into the back of their pickup and we drove slowly along the Athi river looking for animals. Zebra raced ahead of the car, and an old bull buffalo got up arthritically from his spot in the shade and lumbered deeper into the bush. Then a group of waterbuck appeared and the car stopped. I was looking through my heavy old binoculars at the male, a rich-brown, shaggy, stately animal with dark eyes, a heart-shaped nose and a fine set of ridged horns like arms raised in surrender. I was still focusing on him when there was an ear-splitting bang, and the waterbuck jerked back, staggered a couple of steps and fell in the dust. I couldn't believe it. The first wild animal I had got a good close-up look at and I had seen it die. Numb with horror and incomprehension, I followed the man and his boys, and we all stood around the body, so suddenly robbed of its grace and perfection. A small stream of blood trickled into a crack in the black earth.

'Are we going to take some meat home?' asked one of the boys. 'No,' came his father's reply, 'waterbuck meat tastes terrible.' With that the man turned and walked back to the car, and after giving the buck a couple of kicks his sons followed. I was bursting to blurt out 'Then why did you shoot it?' but I was too scared and shaken, and somehow I felt I knew the answer anyway. The boys did not appear to be upset by this pointless killing, and I decided right then that these would not be my friends and I looked even closer to home.

Our staff were WaKamba, a tribe renowned as hunters and better companions no boy could wish for. Musau, the houseboy – as they were called back then – was young and unschooled, and always supplementing his diet with doves that he caught in a variety of traps and snares. Mubinzu, our old cook, took down bigger game when off duty. The WaKamba were sworn enemies of the Maasai, the warrior pastoralists who have become icons for the old Africa. The railway line across the plains that had brought us to our new home was the official boundary line between the Maasai and Kamba tribal reserves, a line respected only by the train. Two or three times a year Musau would shuck his *kanzu*, the white robe that was worn in the house, and Mubinzu would whip off his apron and, wearing just loincloths, they would rush off with their bows and arrows to join the latest scrap with the Maasai somewhere along that borderline. They would return a couple of days later, dishevelled and grinning but unharmed, leading to some doubts about their stories of ferocious hand-to-hand combat.

Every afternoon, when Musau had no more duties until evening, we would walk two miles across the plains, scattering gazelle and zebra ahead of us, to the Athi river, where we would sit in the shade of an acacia and, with simple stick-and-string fishing equipment, would catch fat tilapia that we grilled on a driftwood fire. Musau always brought a small piece of posho, the maize meal that was his staple diet. He would put some on open ground and build a small tepee of twigs around it with a couple of gaps into which he placed snares made of wildebeest tail hairs. Those bird traps were deadly and most days we would have a roasted dove to go with our fish. If I close my eyes I can relive those afternoons now: the burned skin and the burst of hot bloody juice from the underdone dove breast, the smell of the mudbanks and the low calls of mourning doves in the acacias. From Musau I also learned how to collect the red sticky berries from a mistletoe-like tree parasite to make birdlime. Spreading it thinly on twigs near their feeding spots, I would catch small birds, but only so that I could briefly hold them in my hand to take in the tiny details I could never see otherwise. Close up, their beauty was startling. I caught cordon blue finches, cobalt with vivid red earmuffs, and examined the tiny metallic blue flecks in the scarlet bib of a sunbird, the yellow ring round the eye of the firefinch, and the tiny white spots that made it look like a little feathered strawberry. After close inspection I would clean the lime from their feet and, with their tiny hearts pounding, I opened my fingers to watch them go.

I also went out many times with our old cook, Mubinzu, whose favourite way of procuring food was with the bow trap. Mubinzu did not like to set snares, as he said the snared animal called out in distress, and this could attract a leopard or jackal that would come in and take the prize. The bow trap usually killed instantly. The bow was fixed to a couple of stakes about a foot above the ground. The string was pulled back and held with a clever catch, which would release the arrow when triggered by a string fixed across a game path or the latrine of a dik-dik. These slender, cat-sized, twig-legged antelope come regularly to a spot to deposit their dung, and setting any kind of trap there is a safe bet. Mubinzu told me the legend of how a dik-dik had once stumbled over a ball of elephant dung and was so annoyed that he called all the dik-diks together to pile their droppings in one spot, hoping to create a heap large enough to trip up that elephant.

Although I hated the thought of killing such a lovely creature as a dik-dik, Mubinzu was doing it for food. In wartime England we had all known hunger, so I came to accept this kind of death, much as I accepted the daily slaughter at the factory. Then, one day I was out with Mubinzu when he suddenly shot forward into a patch of bush and seized a young duiker, an antelope slightly larger than a dik-dik, which he had spotted hiding there. It struggled violently, but with a quick movement he broke both its front legs. I was horrified and furious, so much so that I actually booted him on the shins as I tearfully cursed his cruelty. 'This can feed my whole family – now they can eat like you do on a Sunday,' he explained, 'but it has sharp feet. If it kicks me and I drop it, it will be gone. Now it cannot escape.' Seeing my concern, and knowing that he could not now lose his prize, he just as swiftly broke the duiker's neck. Although as traumatised by this incident as I had been by the shooting of the waterbuck, I could understand Mubinzu's logic, whereas I would never comprehend that pointless killing. It took me a long time to work it out, but I eventually apologised to our old cook for kicking him.

It seems that even the most pacifist of parents will one day see their son – who enjoys his art and dancing classes – pick up a stick, point it and go 'pow'. It's an atavistic piece of behaviour that appears to be hard-wired into boys, and somehow unaffected by the way they are raised. Despite these early upsets, eventually the time came for me to want to hunt too, so Musau made me a catapult out of a forked stick, some strips of inner tube and a soft piece of goat skin to hold the missile. In readiness I filled my pockets with small round pebbles from the river and decided that there were only a couple of acceptable targets: the mousebirds, known as mousies, and bulbuls, which spent most of their days gorging on the mulberries and guavas in our garden or hollowing out our pawpaws. These vandals were fair game. After many failed stalks and shots, my best chance for a kill came one evening when in the fading dusk I saw the outline of a fat-looking bird that I was pretty sure was a bulbul. Steadying my nerves I let fly. Tragically, the reason my target had looked rather plump was because it was actually a pair of firefinches – a beautiful little brick-red, white-speckled male and his fawn-coloured mate – who had been roosting huddled together for warmth. My rock was big enough to kill both and they tumbled to the ground at my feet, fluttered briefly and were still. They were daily visitors to my bird table. I knew them! Possibly had even held one! Dropping to my knees, I sobbed my heart out and was still there half an hour later when Musau came to call me for dinner.

It would be nice to report that after that trauma I never used my catty again, but not many little boys are made that way. There is undoubtedly a deep ancestral desire and satisfaction in youth for the hunt, to hunt as your forebears did, with a rock, then a club, a spear, an arrow and then with a gun. Man is still a hunter, still a searcher for meat for his belly and his family. Sad as I was about the firefinches, in fact I now yearned for a gun. I felt that mousies and bulbuls were still legitimate

targets, and argued that reducing their numbers would be beneficial and more humane if I used a gun.

Shortly after arriving in Kenya my sister and I had been
enrolled at a school in Nairobi, and were ferried the twenty
miles back and forth every day in the old Austin belonging to
Mr Patel, the owner of Athi River's only shop. In the afternoon
we were collected from outside May and Co., Nairobi's biggest
sporting store, where every day I gazed longingly through the
window at a Slazenger .22 rifle priced at five pounds with a
hundred rounds of ammunition.

My parents said an emphatic no to the .22, and my eleventh
birthday passed with dreams unfulfilled, but the next Christmas
there was a .177 airgun under the tree. I got to it well before
the folks were up. Anyone using a gun for the first time should
have someone familiar with weapons give him or her a thorough
introduction to handling and safety. I had no idea how to use

a gun and my parents – who
were barely qualified to help
me themselves – were still
in bed. But on that Christmas
morning I was burning to use
it and thought it must be
pretty elementary, so I
successfully fired my first
blast without putting a pellet
in. The sound got my mother
out of bed very rapidly, while
I continued to experiment. I
cocked the gun again –
folding the barrel down to
compress the very powerful
spring – then looked down
the hole that led back into
the compression chamber. I

LOUTH
582503

was now holding the gun with the hinge close to my face and the folded barrel hanging down like an extension of my nose. I wish I could say that this was enough to satisfy my curiosity, but no, further research was needed, so I pulled the trigger. The folded barrel slammed back into position with tremendous force, hitting me squarely between the eyes, with the front sight digging a hole in my forehead. My mother rushed in to find me unconscious and bleeding, gun in hand. 'Alan's shot himself,' she screamed. After that, watching my sister unwrap her dolls was a comparatively serene affair.

To keep up with the demands of the 'Bully for Britain' campaign my father's factory slaughtered cattle from all over Kenya and Tanganyika (now called Tanzania), and had developed a widespread network of cattle buyers, stock routes and holding grounds. One day I was unexpectedly invited to accompany my dad on one of the tours he made around the buyers in Tanganyika. One of these men was a flamboyant Texan cattleman called Oscar Dahl, who wore cowboy boots and a Stetson, carried a stock whip, and had the huge horns of an Ankole bull mounted on the bonnet of his Ford Mercury wagon. Dahl knew that he was a magnet to kids and was generous with his time, talking of his adventures and teaching me to swing and crack his long whip. As we were saying goodbye to him he suddenly asked my dad, 'Why don't you leave the boy at Longido? I'll give him my whip and he can join the drive and walk to Namanga.' Seeing my father hesitate, Dahl then said, 'It's only twenty-five miles and I'll keep an eye on him.' So Dad acquiesced to his leathery employee, who later that day took me in his wagon to join the cattle drive at Longido, a spectacular eight-thousand-foot mountain.

Back then, Longido was the official point of exit and entry between Kenya and Tanganyika. Not a soul lived there and, as there were no government officers to monitor the border crossing, officialdom was represented by a permanently raised barrier and an unmanned sentry box with a dog-eared book on the shelf to record the details of travellers crossing back and forth. As Oscar delighted in pointing out, it was amazing how often Charlie Chaplin, Adolf Hitler, Darwin and Einstein had passed that way, sometimes all in the same vehicle. We signed out of Tanganyika as Donald Duck and Tarzan and, since that first experience, borders and bureaucracy have never meant much to me, particularly in Africa, where the lines drawn on inaccurate maps by colonial powers divided tribes and king-doms, and sowed the seeds of many of today's conflicts.

Dahl drove us to a spring at the western base of Longido mountain, where I joined a group of about a dozen Maasai. They were camped with a herd of around five hundred cattle that they had slowly marched up from Babati, a cattle market a hundred and sixty miles away in Tanganyika. The group seemed very young and bristled with weapons, each carrying a spear, a long, sheathed knife and a *rungu* – a short, heavy throwing stick with a large round knob on the end. Oscar gave me his sleeping bag, his stock whip, water bottle and some dried meat called biltong, and said he would see me in Namanga. I watched apprehensively as his lanky figure strode away, swallowed hard and turned to face my new companions.

By now I could speak enough KiSwahili to communicate with a couple of the Maasai herders and the next four days were a wonderful adventure. Every morning the warriors would pull a bull out of the herd and, from a couple of feet away, fire a short arrow into its jugular vein. The blood was caught in a gourd and the tiny wound sealed with a plaster of dung. It was then whisked with a stick, which collected a large fibrous clot, and this gory lollipop was eagerly eaten by whoever's turn it was that day to get this special treat. The remaining frothy blood was mixed with milk and offered to me. They all watched with great interest – clearly, Maasai etiquette demanded that I drink up – so I took a swig of the warm foaming fluid. It wasn't as bad as I'd feared, with a meaty, salty taste, but the smell brought back memories of sloshing through welly-deep blood on the killing floor of my father's factory. Every morning, after satisfying the warriors with just one swallow, I was able to move on to the maize-meal porridge and milk.

We were up at first light to wonderful bird calls. Red and yellow birds duetting madly on an anthill, and a deep booming from some black turkey-sized fowl with pickaxe beaks. I had no idea what they were and longed for a guidebook. The dry bush country has the finest collection of colourful and intriguing birds of any habitat I have known, and I blessed my heavy binoculars as they thumped uncomfortably against my chest. My memories of even the most vivid illustrations in my British birds book were rapidly fading in the equatorial sun. Every day was filled with so many new sights and sounds. The acacia bushes with their grappling-hook thorns were in full bloom, a multitude of butterflies and irridescent beetles clambered among the froth of creamy flowers, and the air was like breathing powdered honey. As we walked, two species of graceful antelope I had not seen before – the slender, long necked gerenuk and spiral-horned, gunmetal lesser kudu – darted in and out of the thorn bush ahead of us.

We moved slowly, allowing the cattle to keep grazing so they didn't lose weight. Late each afternoon we would arrive at a ready-built, thorn-fenced boma where the cattle would spend the night, kept relatively safe from lions, if not from marauding local Maasai who regularly attacked the drives and made off with as many cattle as they could. My clothes, hair and nostrils were full of dust and the smell of the drive. When a strap on my very English Clarks' sandals broke, one of the men sewed it up using a porcupine quill as a needle and a thin strip of hide. In the evenings, after maize porridge and a mug of tea, I fell asleep quickly on the ground near the fire while the men, wrapped only in their lightweight loincloths, spread themselves protectively around me. Each night-stop was close to water, either a natural spring, or holes dug in a sandy river bed where I could have a rudimentary wash. One night the roaring of several lions announced that they were testing our defences. I was told to throw more wood on the fire as the Maasai noisily threw burning branches out into the night, and the lions decided against a raid. The next morning we looked for their tracks and nearby found the large three-toed imprints of a rhino, which the Maasai said were common and often panicked the cattle with their snorting and short-sighted charges.

On the fourth afternoon, as I said goodbye to my guardians at Namanga, I realised that I felt a changed and more confident person. Only much later in life did I recognise and value that initiation. Those young warriors, whose tribal customs practise and respect such rites of passage, which have been mostly lost in Western culture, would have clearly understood. At Namanga there was nothing but the Namanga Hotel, a group of thatched rondavels nestled under giant fig and acacia trees on the banks of a river that flowed cool and clear from the nearby hills. The hotel's entrance was framed by a large pair of elephant tusks and all the doorstops were rhino horns that had been found in the surrounding bush. Oscar had made sure that there would be lots of hot bathwater for my room and I enjoyed a therapeutic wallow. My father came to collect me the following day and we stopped for a picnic under some huge acacias with a fine view of Kilimanjaro. I remember Dad took a picture of me next to a termite mound, where I found the shed skin of a large snake draped over the top. Then he had handed me a sandwich, opened a beer for himself and poured a little into a mug for me to celebrate my trek. This is a rare memory of any kind of bonding or shared experience with my old man.

With Oscar Dahl fighting my corner I was hoping for more and longer trips with the herdsmen. Maybe next time I could join the route down from northern Kenya? But soon after that trip my father was promoted and we moved to a remote factory in the bush, at West Nicholson in Southern Rhodesia (now Zimbabwe) – this time I left the toboggan behind. West Nicholson was as wild as Athi River, and laid out in a similar pattern with the factory and a row of identical houses along a dusty road that ended at the club, with two tennis courts and a green and scummy swimming pool. The whole village was strung out on the banks of a wide sand river, the Umzingwane, where animals came through the dense forest of figs and tamarinds to dig in the sand for water.

Echoing my arrival in Kenya, at West Nick I was invited to join a group of youngsters on a 'torch safari', which sounded wonderful until I discovered – too late – what it really involved. Powerful torches were used to find the shining eyes of small nocturnal animals. This I found fascinating, but the kids then approached within a few yards of these blinded creatures and took turns to blow them away with a shotgun. They got their thrill from firing the powerful 12-bore with its spurt of flame and heavy kick: the animals simply provided a more interesting target than a tin can, and one that evoked no more compassion. That night they shot two small, spotted carnivores – a genet and a civet – and also a porcupine, who came snuffling, blinded and unaware, to within a few feet of the shooter. I couldn't opt out because I didn't have my own torch and would not have found my way home. Once again I had seen a new animal for the first time, only to watch it being not just killed, but almost shredded in a hail of lead.

I rejected the company of the young hunters, but there were no Africans to befriend and learn from in West Nick, for our staff there were all Shona, an agricultural tribe, which at the time did not hold the same promise of excitement as my WaKamba or Maasai friends. So I began to spend more and more time on my own, following the shining threads of water that meandered along the wide golden sands of the Umzingwane. Taking some bananas and biltong, and drinking from the river, I carried my first African bird book, the heavy *Robert's Birds of South Africa* that had been my birthday present. Now I could identify the tiny, fearless spotted owl with the pendulum tail that I had met in the deep shade of the forest. With its large yellow eyes and strange clockwork movements, it had seemed a frightening little creature from the world of darkness. It was a relief to thumb through *Robert's* and

identify it as the pearl spotted owlet – not the occult creature I had imagined. Similarly, the crested flycatcher with the trailing white tail and the eagle with the floppy, black crest now had names and became familiar. I was alone, but I was happy.

Life in West Nicholson, as with many small expatriate outposts of the old Empire, revolved around the club. My father only went on special occasions – when someone was leaving, or there was a prize-giving or similar – sitting around talking and drinking was never his scene. My mother tried it, but she found the patrons very parochial and there was no one to joust with verbally or share her sense of humour. She hated Southern Rhodesia, and even more so now that my sister and I were sent to boarding school a hundred miles away. In fact, she disliked her life there so much that she wanted to take us back to Kenya, persuading my father that he should apply for a transfer while she took us on ahead. I guessed that my parents weren't getting on too well and certainly we hardly ever saw my dad, who seemed to live at the factory, but the upside was that a year later my mother, sister and I headed back to Mombasa on a little British India Line ship.

In every port along the way the lascars – Indian seamen – provided, then tirelessly untangled, fishing lines for all the children on board, and helped us to bait our hooks with bright yellow curry-flavoured dough. I hooked a remora, the sucker fish that attach themselves to manta rays and whale sharks with an oval sucker located on top of their head.

The sailor who had helped me bait my hook, and now helped to unhook the fish, was really excited. 'Very good luck for you, little sahib, very good luck.' When I asked him why it made me lucky, he replied, 'This small fish is sticking himself to great monsters of the deep. Giant rays and sharks. Then he doesn't have to swim any more and just lives on scraps from his big friends. Now you have caught him you will be like that. Getting plenty scraps from your big friends. Very lucky, sahib!' I thanked him for the story, but somehow that wasn't the future I had in mind.

Now back in Kenya we no longer lived out in the bush. Since my mother wanted to find a job, and my sister and I were enrolled at secondary school, it was easier to live near Nairobi. We moved into a mud-and-thatch house on the edge of the forest at Karen, a sparsely inhabited area some ten miles out of town. I was at the Prince of Wales school – the 'Prinso' – while my sister Jackie was at that potting shed of the Kenya rose known to all as 'the Boma', the full nickname being Heifer Boma, after the thorn-fenced enclosure where the young unmated cows are kept. As we lived across town and had to travel to school by bus, I was allowed to leave shortly after lunch, thereby avoiding any sporting activities, which suited me just fine. Unbeknown to the school, I soon managed to organise a lift with a neighbour, which meant that I was home early enough to go straight out into the surrounding Karen forest, where I spent the afternoons watching a whole host of new forest birds – scarlet and emerald narina trogons, red-winged touracos and nesting crowned eagles. In the middle of the forest stood a fifty-foot water tower that, complete with ladder, provided a perfect treetop viewing platform. I could watch hooded vultures and orioles nesting by day and would take a torch and a sandwich and spy on bushbabies and eagle owls at night. The tower had the added virtue that on hot days I could slip through the inspection cover to cool off in Karen's water supply.

The boys at our school came mainly from farming or outdoorsy backgrounds, the sons of professional hunters and animal trappers, ranchers, game wardens and forest or fisheries officers. Many of them considered hunting a good way to spend their free time and, although I still spread fear among the mousebird population with my trusty airgun, I was losing interest in shooting and found several boys who shared my less predatory feelings about nature. Chief among these was my friend Nick Forbes-Watson, a boarder whom my mother invited to stay most weekends. Nick shared my interest in birds and snakes, and was already an accomplished ornithologist. He was a superb tree and rock climber, and had started an egg collection that in later years would become important enough to be accepted by Nairobi's Coryndon Museum. I soon built up a good collection of snakes, which I caught in the forest or along the nearby river. My mother encouraged this hobby, as long as the snakes were harmless. Fortunately Mum didn't know her asp from her emerald snake, and without telling any real fibs I worked on the principle that unless you get bitten, all snakes really are quite harmless. My collection, which soon included deadly species like the puff adder and boomslang, was given the stamp of good housekeeping by dear old Mum whose modest abode was fast becoming a small zoo as I accumulated various orphan animals – bushbabies, a genet cat, duiker antelope and a baboon called Bimbo, who became part of the family. Mum's tolerance for these lodgers was remarkable, but she could not help me with natural history and I longed for someone who could teach me more.

In my school library I had discovered a book called *Great Northern*, about a group of English children who had discovered a nesting great northern diver bird, which they had helped to protect. I showed the book to Nick, and he became very excited on seeing that it was dedicated to M.E.W. North, the name of the District Commissioner for Nick's local area. The DC was the senior government representative in any given area and on the pretext of wanting him to sign our book we paid him a visit. Myles North could not have been more welcoming. An idiosyncratic bachelor, with precise and plummy diction, and a stammering but amusing delivery, Myles's whole life was birds and he tried to arrange his postings so as to spend time in areas that would provide the best birding opportunities. With a call for tea and biscuits Myles gathered us in like a broody hen. He had just started recording African birdsong with equipment provided by Cornell University. He explained that the operation required the help of two people to carry the equipment as they sneaked up as close as possible to a singing bird. He had just described the abilities of a pair of experienced mousy hunters like us and by the time we had finished the digestives Nick and I had been offered a wonderful way to spend our holidays, and I had found a mentor.

Myles had a small Austin station wagon and packed in foam rubber in the back was a huge Ampex tape recorder, about thirty inches square and eighteen deep. Lord knows what it weighed, but because it could be lifted by two men it was considered to be one of the first 'portable' recorders. It required a hundred and ten volts of power, which was provided by a rotary converter working off the car battery. This was a very inefficient process that often left us with a flat battery requiring a push to get the car started. Fortunately the car itself did not weigh much more than the tape recorder. Now on weekends, instead of setting out with bikes and buns, we would drive out into wild country with packed lunches and a thermos of tea, stopping frequently to listen for bird calls, and when Myles identified a target Nick and I would take turns with the equipment. One of us would slowly unwind a reel of heavy cable while the other carried a tripod to which was attached a four-foot-diameter parabolic reflector, which focused the sound on to the microphone. We had to get as close as we could without disturbing the singing bird and aim the parabola at it as accurately as possible, while Myles monitored the sound through headphones back in the car and captured it on tape. I vividly remember catching the calls of a group of wood hoopoes, their long black-and-white tails waving in unison, chuckling and giggling manically as if one of them had just told a really good joke. Another day it was the calls of nesting marabou storks, moo'ing, hissing and clattering their pickaxe beaks like castanets.

Before long, Myles had engineered a transfer to Voi, a hundred and fifty miles away down the road to the coast, in the dry thorn country on the edge of the great Tsavo National Park. Voi consisted of a railway station, a Somali butcher, an Asian shop with a petrol pump and the District Commissioner's office – and there was also always a small group of elephants somewhere between these scattered buildings. Myles caused great amusement and embarrassment when he gave his welcoming address to the locals, including a smartly turned-out honour guard of National Park rangers. He droned on with the usual pleasantries, how he had heard tell of the friendly people and progressive attitudes of his new principality, but then he excitedly blurted out that he was 'particularly pleased to be in Voi because there is a little red-and-yellow bird here that in the morning sits on a termite mound and goes "kerwootertytok kerwootertytok" and waggles its tail . . .' And here, along with the high-pitched song, he gave an energetic imitation of the vigorous rump-wagging that accompanies the barbet's display. All down the rigid line of well-disciplined rangers eyes were shooting side-ways, lips pursed tight, trying not to smile, and you could almost hear them muttering, 'Oh God, what have they sent us this time?'

Nick and I spent two school holidays staying with Myles at Voi. They were among the happiest and most free times of my life. With string and tape I attached my Box Brownie camera to one eyepiece of my binoculars, and by sighting through the other side I started to take my first photographs of wildlife. We spent the days wandering that vast, wild, unpeopled country, following the Voi river through the hot, dry bush, or climbing high into the cool forests of the nearby Taita hills, and would come home in the evening to showers and good food, but best of all to discussion about what we had seen that day and what Myles could add to our observations. Like many colonial children, we received much of our education in the school holidays.

As Myles's guests, Nick and I were also invited to a memorable dinner at the fabulous hilltop Moorish castle of Colonel 'Grogs' Grogan, the man who fifty years earlier had spent three years walking the length of Africa, from Cape to Cairo, to prove his worth to the father of his fiancée. He returned with her to Africa and became a founding father of colonial Kenya, where he built the deep-water port at Mombasa, one of Kenya's leading hotels and the children's hospital Gertrude's Garden – named after the girl who had waited for him as he made that epic walk. As genet cats patrolled outside the huge arched windows, catching moths and beetles, this leonine and historic gentleman, then in his late seventies, told us youngsters tales from that walk: of finding the skeleton of a gorilla in the Congo, of stumbling upon cannibal feasts and a head with a spoon left sticking in the brains, of adventures and wild animals beyond my imagination. These stories were disrupted at intervals by the fights that broke out between his large pack of Pekinese dogs, whereupon several soft-footed servants in robes and red fez would sweep in, scoop up a snarling, bug-eyed pooch under each arm and silently sweep out again.

School was suffocating me. My father, who was still living down in Southern Rhodesia but had a hand in my future, wanted me to aim for university, as did my biology teacher, but I wanted out. Although I lived at the world's best address for a naturalist, I desperately wanted to visit the Amazon. The Congo would have provided a better rainforest habitat right next door to us in Kenya, but I had been influenced by a series of books about the adventures of *Bomba the Jungle Boy*, which took place in the forests of Brazil. Every few days I picked up a new book from the public library in Nairobi, and each was full of jararaca and fer-de-lance snakes, of jaguars, peccaries, anacondas and Indians with poisoned-dart blowpipes. It all seemed so much more exciting than home and I just ached to get out of school and to South America. Now sixteen, and having just scraped through my exams, I had a modicum of education, so finally my father, who had done pretty well on a lot less himself, gave up the fight. He decided that I could leave school on condition that I took a job with his firm, Liebig's. This bit of nepotism suited my plans well. It got me out of school and, as the meat-packing company had extensive cattle interests in South America, I felt sure I could soon transfer to the Amazon.

2

On the Road

My first job was as a travelling salesman for my father's employers, Liebig's. It sounded awful but in fact turned out to be wonderful. My beat was the whole of Kenya, Uganda and Tanganyika, and I was given an Opel truck full of samples of the canned meat, fruit and vegetables that the company produced. I could dress informally and work out my own itinerary, so I set off with an order book to visit each corner of East Africa.

I hit every one-store town there was, where the inevitably Asian owner would know exactly what he needed. I would be given a cup of masala tea, some sweet and sticky confectionery and, with zero salesmanship, would receive an order in quick time, which I telegrammed to head office whenever I came across a post office. After a while I worked out a way to maximise my time in the most interesting areas. I would race through the more peopled stretches of my territory at high speed and get ahead of my schedule, so I could spend a couple of days birding or snake catching in the more remote and rewarding spots. So I wouldn't need the traveller's inns where the company expected me to stay, I threw a camp chair and a mattress into the back of the truck, and handed out fewer samples of corned beef, tongue, tinned mangos and grapefruit – now I could camp out. One evening in southern Tanganyika I drove a mile or two off the road and parked below a huge

baobab tree with a view of a vulture's nest high in a neighbouring borassus palm. There I sat with my binoculars, content-
edly eating my company's products and sipping a warm beer,
thinking, 'Hey, I've got this job organised,' when, as the sun
began to go down, I was treated to a wonderfully unexpected
spectacle. Scores of lovebirds – sparrow-sized, bright green
parrots, with orange heads, scarlet beaks and a fleshy white
ring round their eyes – were coming in to roost. They came in
as a chattering spray of emerald bullets, to perch, like a line
of brilliant musical notes, on the curving spine of a palm frond,
which gleamed gold in the warm evening light. After a couple

of minutes of colourful and noisy bickering, the birds all shuffled sideways towards the large nest of the vulture, then
burrowed in and disappeared, to sleep safely inside that deep
pile of sticks. The vulture, about as far as possible from the
lovebirds on a scale of beauty, sat peering tolerantly over the
edge of its nest as this horde of squatters moved in, just inches
below its heavy beak. I lay awake that night under the stars
and every time there was a wave of chatter from the restless
lovebirds I got to wondering what I could do with my life that
would enable me to spend more time this way.

In those eighteen months I saw as much of East Africa as I
would in the rest of my life: Lake Rukwa, the gold-rush ghost
towns and the vast baobab woodlands of southern Tanganyika,
the lakes and desert outposts of Kenya, the Mountains of the
Moon and the volcanoes of western Uganda – and everything
in between. The Amazon didn't have a chance; I now knew
where I belonged. I also knew that seeing all that wildness and
beauty as a tinned-food salesman was not good enough. I needed
to find a way to be more involved, to be out there doing something, something unnamed that was waiting for me to make it
happen. I wasn't looking forward to telling my father I wanted
to leave the company, knowing that he would point out the
many advantages of sticking with a safe job and my complete
lack of credentials for anything more promising. I couldn't blame
him – his own life had been such that leaving a good job without
a better prospect was unthinkable. But while I was still plucking
up courage to start that conversation with my father I received
an official-looking envelope. I was pushing eighteen and they
were call-up papers for my National Service with the Kenya
Regiment, giving me an honourable way of getting off the hook.

Kenya was then in the midst of a State of Emergency – a
euphemism for war – which had been declared by the colonial
government in 1952 to counter the Mau Mau uprising. Call the
Mau Mau what you will – terrorists or freedom fighters – but

this was an extremely savage revolt primarily by the Kikuyu tribe, aimed at removing the white man and the colonial government from Kenya, whatever the cost in blood. They believed the white settlers would flee, closely followed by the government. The victorious Kikuyu tribe would take over the settlers' deserted land, then the country and the government. There were troops and roadblocks in Nairobi, many whites wore pistols or carried guns and remote country farms were heavily fortified against attack.

In the early years of the twentieth century the land in the Kiambu area, some fifteen miles north of Nairobi, had been unoccupied except by wild animals and was seen as ideal for the white settlers who were arriving at the time. Unbeknown to the officials responsible for these decisions, just decades earlier much of this area had been inhabited by large numbers of the Kikuyu tribe who had been decimated by a series of disasters: a devastating drought, a major locust invasion, an outbreak of the cattle-killing disease rinderpest and smallpox, which had reduced the population by as much as fifty per cent. Most of the survivors had moved back to the Kikuyu heartland in the Muranga and Nyeri areas further north. But they had left because of bad times and although their fields quickly reverted to bush and filled with wild animals, they had always considered the move temporary: they would move back when conditions improved. Their later discovery that their 'empty' land now belonged to white farmers was one of the key events that led to an increasing resentment of whites and the colonial government.

The Kikuyu were the most advanced tribe in Kenya, and in the late forties there were many young Kikuyu with mission-school educations and high expectations, including a number who had attended British universities. Just one generation back they would have been warriors, respected defenders of their tribe, territory and customs. No longer comfortable with their own tribe nor qualified enough to find the work of which they dreamed – without heritage or hope – they gravitated towards the towns and the firebrand orators who were preaching revolt. These deracinated young men were a fertile breeding ground for the movement and they began recruiting others by means of oathing ceremonies. In the recent past, the oath had been an extremely powerful means of maintaining tribal discipline and traditions, and was an absolutely binding instrument. For the purposes of the Mau Mau revolt, this oathing system was perverted into a series of increasingly obscene and degrading acts with escalating demands so that those who took the oaths, whether willingly or under duress, were bound to commit the most depraved atrocities.

Our army training took place near Nakuru, a one-donkey town on the edge of a flamingo-haunted soda lake on the floor of the Rift Valley, the two-thousand-foot-deep gash that runs down almost the entire length of Africa. The many miles I had walked, and the tree and rock climbing skills that I'd polished while egg collecting, now came into play on the cross-country obstacle course. While explosions threw mud in the air and Bren-gun fire cracked and thumped overhead, I slid down ropes, scampered up cliffs and waterfalls, up, down, over and through everything that the trainers could think of to slow us down. I was good at this stuff, so they decided to keep me on as an assistant physical training instructor rather than send me out on active duty. It would be only for the duration of the next course, but I itched to get out there. Meanwhile I was climbing ropes, vaulting horses and walking backwards for several miles every morning, so I could keep an eye on the straggling line of gasping recruits. Frustrating as it was, the six weeks passed quickly, and at the end of it I was as fit as I could be and was moved on attachment to a British regiment.

The Emergency had been going on for some four years, and most of the fighting was over. The action now was really a mopping-up operation, searching for the few remaining Mau Mau gone into hiding – dwindling but still-sharp needles in the haystack forests that covered the Aberdares and Mount Kenya. Any Mau Mau still alive in those forests was a hard-core survivor and would take some finding. My role with the British Rifle Brigade Regiment was as a tracker master, where I would be in charge of, and interpreter for, the African trackers attached to the regiment. Also attached to us was an ex-Mau Mau, known only as Mbugua, who had come over to our side, as had thousands of other Kikuyus, now sick of the way the uprising was blighting their lives. Fortyish, powerfully built and wiser than his years, he had a good sense of humour, especially regarding the hardships of the life he had led. He had spent several years trapping animals to feed the gangs in the forest and with my interest in wildlife we soon became firm friends.

One day he offered to show us the hollow trees and caves along the length of the Aberdares that were used as drop-off points for food and information, places where it might be useful to set a waiting ambush. I convinced the officers that a reconnaissance trip of such uncertain length could not be undertaken by a full patrol, and that Mbugua and I should go alone. As there were only two of us he had to be armed, but although he had only come over to our side a few months earlier, I felt fairly confident I could trust him. It was a surprising but common feature of that time that men who had been bound by the most degrading of oaths, and were sworn to kill whites, could change sides, take a powerful absolving oath and quickly become an ally and friend. They had thought we were feeble, that we would abandon our pedestals and run, but we didn't, we took the fight to them out in the forests. We felt we were superior but once in those forests we realised that we definitely were not, and they had the drop on us out there. Based on these realisations, out of the conflict grew a mutual respect and understanding that was real and lasting, and many ex-Mau Mau moved on to become rangers in the Aberdare and Mount Kenya National Parks, helping to defend the forests they knew so well.

Mbugua and I set off from our base at South Kinangop, at the southern end of the Aberdares, the sixty-mile-long north-to-south volcanic range that forms the eastern wall of the great Rift Valley. We carried just groundsheets, the makings for tea, a few tins of corned beef and biscuits, and that wonderful invention: self-heating tins of soup and cocoa. There was more fresh water up there than we needed and Mbugua assured me he would catch us plenty of meat. Our journey started at around seven thousand feet, through a dry forest of silvery leafed crotons, gnarled olives, tall cedars and feathery podocarpus. The high forests of the Aberdares are stunningly beautiful. Variations in temperature and rainfall at different altitudes produce belts or zones of differing vegetation and after climbing a couple of thousand feet we entered a totally different world – the bamboo belt.

Here visibility is limited to yards by the closely ranked stems of wrist-thick, thirty-foot-tall giant grass. The stems rattle and click in the wind, which moans eerily over the hollow ends of dead stems. The only way through is along game trails, but walking them you run the constant danger of meeting the architects of those trails, with no easy way to escape through the densely packed stalks. Six to eight feet above the ground the green outer layer of the bamboos was scraped off and stained red by the rough-skinned passage of elephant, and we had to make several long detours around the herds that we met en route. Here we also found the tracks of the bongo, one of Africa's rarest and most beautiful antelopes, shy chestnut-coloured animals the size of a cow with some twelve narrow vertical white stripes and spiral horns. Several times we glimpsed them as they raced away, their horns clattering on the bamboos. Above the bamboo zone, at about ten thousand feet, the country opens out to glades and great stands of one of our most beautiful trees: hagenias, their huge trunks covered with flaky red bark and their low, heavy, horizontally spreading branches carpeted in thick rugs of bright green moss and hung with long silvery strings of beard moss and

reddish inflorescences. Very little vegetation grows below hagenias so it is possible to see a good distance ahead and move
quietly on a deep carpet of large lemon-and-rust-coloured leaves
– damp, soft and silent underfoot. Climbing through them to ten
or eleven thousand feet you come to the heath zone, where giant
heather and St John's wort – small shrubs in more temperate
climates – grow to twenty-foot trees. Among them the crotalaria
bushes attract dazzling sunbirds – double collared, tacazze, golden
winged and malachite – that flash among the yellow leguminous
flowers, brilliantly living up to their extravagant names. Higher
still and you break out into sunshine and the alpine zone.
Tussocky grasslands dotted with lobelias, shaggy six-foot columns
covered with tiny blue flowers, and senecios, twenty-foot
branching trees with giant emerald cabbages at the end of each
limb that sprout spikes of bright yellow flowers: primitive-looking
plants that only survive at these altitudes by tightly folding their
leaf rosettes at night and exuding a mucilaginous anti-freeze that
protects the leaf buds. Every morning we would easily spot the
six-inch-long Hohnel's chameleons, basking exposed on the sunny
side of heather bushes, coloured an eye-catching matte black to
absorb as much heat as possible. An hour later and they would
have warmed up, changed to camouflage colours and be invisible
for the rest of the day.

Hiking across the grain of the mountains meant a lot of ups
and downs from ridge top to valley floor, where we would
inevitably have to cross an icy river before heading up again.
Mbugua could name every stream, confirmed each time by my
large-scale maps. Periodically he would stop and look at me
questioningly – a signal for me to pick out the tree that had been
used by the gangs as a drop-off site. One tree I picked out very
quickly, because of old scratch marks that had torn the moss
from around a hole in the trunk. Mbugua explained that a leopard
had tried to get into the tree, going after an antelope leg that
Mbugua had secreted there. Telling me the story reminded him

that we were running low on food and that night he easily snared a duiker antelope, which we made last for the two days it took us to skirt round the west of the great domed rock peak known as the Elephant, past Fey's peak and the headwaters of the Tulasha, and on to the Wanjohi river. Back in the twenties and thirties, the Wanjohi was said to run not with water, but vintage champagne. From the banks of the river we could look out over the infamous Happy Valley, where a handful of rich and dissolute aristocrats, bored out of their empty heads by life in paradise, got into drugs, sex, booze and murder, and gave the rest of Kenya's hard-up and hard-working settlers a bad name that still resonates with the uninformed today.

Turning west and uphill, we made a numbingly cold crossing of the spine of the Aberdares at some thirteen thousand feet. The altitude gave me a splitting headache, leaving me unable to appreciate the fields of scarlet gladioli and the stunning views of the snaggled lichen-covered lava spikes called 'The Dragon's Teeth'. That night on Sattima, the high point of the Aberdare range, our water bottles froze, but fortified by a roasted francolin, a small game bird plump as a capon, and garnished with bamboo shoots, we made it to the eastern side of the range. Here we crossed the headwaters of the Uaso Nyiro, beginning its journey out of this cool green world to meander north through scorching elephant country before disappearing into the desert's thirsty floor. On the Mwathe river we checked on Mbugua's last drop-off tree and finally emerged from the forest near the well-named Charity Farm, where we drank steaming sweet tea and called for a vehicle to collect us. We had only covered about fifty miles as the crow flies, but the undulating terrain had more than doubled that. We had checked on eighteen drop-off sites, five of which were still in use, which we tagged as worthy of an ambush. It had also been a great wildlife safari and I had absorbed a huge amount of forest lore from Mbugua, knowledge that one day I would put to good use in more peaceful pursuits.

When my National Service ended my worldly goods consisted of my jeep, a sleeping bag and, best of all, my father's old 8mm Bolex camera. Freed from discipline and the cold, wet mountains, my old friend Nick and I celebrated by taking off on a long tour of the national parks and game reserves, sleeping on the ground, living off the land, filming rhinos and elephants chasing the jeep, potting francolin with the airgun, pinching steaks from predators' kills and filling up on bananas. We had a wonderful time and I was determined to find a job in one of these wild places.

The Coryndon Museum in Nairobi was a wonderfully welcoming place for any young naturalist, and when I was back I visited the curator of birds, John Williams, who in turn introduced me to John Pearson, a commercial airline pilot with dreams of becoming a wildlife film-maker. John asked if I would like to help him with a project to film the jacana or lily-trotter, a water bird whose extraordinarily long toes enable it to do just what its name suggests. He needed help on a recce to find the best place to film them, to put up hides and get the birds used to human presence.

We found our spot at Lake Naivasha, a Rift Valley lake just fifty miles from Nairobi. A belt of papyrus a hundred yards out from the shore provided protection from wind and waves off the lake, resulting in tranquil lily-covered lagoons that were the perfect settings for the film. I returned alone a couple of days later and set up a little tent at Hippo Point on a farm owned by the Carnellys. Then in their sixties, the Carnellys were an archetypal colonial couple, living in a magnificent Tudor-style house among giant yellow-barked acacias. I could fish for tilapia, and was allowed to help myself from the large vegetable garden, orchard and dairy, so camping on the shore with views out over the lake was borderline decadent. The jacana is a rich chestnut colour, with a golden breast, white throat and a pale-blue bare shield on its forehead. There were many of them stalking over the lily pads in front of my tent and on my first day I found several of their nests – flimsy, sodden platforms built of weeds with the eggs nestled just above water level. The jacana's eggs are beautifully marked – a hieroglyphic maze of slim dark squiggles covering a gleamingly varnished chocolate-coloured base. I built two hides sitting on gum poles in the eight-foot-deep lagoon and every day moved them slightly closer to a nest. After a week John turned up to start filming but had only two days before he was called back to fly, so he gave me a quick course on his 16mm Bolex and left me to it. The Bolex was a simple wind-up camera with interchangeable lenses, but no way to focus through the lens. It was all guess-work, so with a tape I measured the distances to the nest and other key positions, and made a list for quick reference.

The 8mm home-movie material I had shot of my pets, or of my safari trips with Nick, had given me a good grounding in exposures and the importance of a sturdy tripod, but I soon discovered that I somehow seemed to know everything else instinctively. It just came naturally: the establishing shot followed by close-ups, the zoom in to a point of interest, or out to give context, the composition, lighting and framing, the change of angle from shot to shot to avoid the appearance of jumping towards the subject, the cutaways, the reaction shots. I found I understood the grammar of film, which was just as well as there was no way I was going to be taught it – I had to pick it up as I went along.

By the time John finally returned, I had already filmed the jacanas' courtship and mating, their nest-building and incubation, and close-ups of the eggs hatching. The tiny chicks are like long-legged bumblebees – round, fluffy and beautifully camouflaged. I had also filmed some great sequences of the chicks being picked up by the adult, for when danger is near – a python, hawk or monitor lizard, for example – the male makes a special call, bends forward and partly opens his wings in invitation. The young, two on each side, climb up underneath his wings, wedging themselves into place, and the adult races over the lily pads with the chicks' long toes – twenty

on each side – trailing like pale-blue spaghetti from beneath its tightly shut wings. In one sequence I filmed the male making a long leap across a gap in the pads, where he could only make it by opening his wings to get some lift. As he did so the chicks tumbled into the water, but he swiftly pulled them out with his beak and they climbed back under his wings.

I knew it was the male taking care of all these family duties because while filming the mating I had noticed that the female was much larger than the male. Although the sexes are similarly coloured, this size differentiation made it easy to observe that not only did the male do all the incubation and care for the chicks, but that the females were polyandrous and had one or two other males, all sitting on eggs or with young, while the female's main responsibility was defending their territory. I pointed out this unusual role reversal to John Pearson, but when the great naturalist Peter Scott showed our film on his early BBC programme *Look*, he decided not to mention these findings as they hadn't been written up and presented as formal research. This was the first of many examples of animal behaviour that I filmed over the years, but failed to publish, thereby losing any claim to the discovery. My old mentor Myles North had given me many lectures on the need to take careful notes and to publish any new observations I had made, but apart from brief film notes I had never followed his advice. The same was true of my later filming and observations of the nesting behaviour of flamingos, the inside-the-nest story of breeding hornbills, wild dogs hunting and regurgitating food for their pups and baboons killing antelope and even taking a kill away from cheetahs. All these findings I had made public on film years before they were officially discovered by scientists. But I guess that is part of the African experience – after all, many geographical features across the continent were named after English and European royalty despite being well known to the locals before being officially 'discovered' by the explorers who first recorded them. That said, several of my films have seen long service in the lecture halls of Oxford and Cambridge, Harvard and UCLA, used by academics as renowned as E.O. Wilson and Jonathan Kingdon – tribute enough for me.

Despite his success in having his first film shown on BBC TV, poor John Pearson just couldn't convince his wife that wildlife film-making was a respectable and profitable career, so he finally gave up the struggle and I was out of a job. With only half a dozen wildlife cameramen in the world at that time it wasn't exactly a realistic career choice, and in any case I

would have needed several thousand pounds' worth of equipment – a fortune in 1956 – before I could have offered my
services to . . . whom, exactly? Then, one day, coming out of
the Standard Bank in Nairobi where my mother worked and
where I had just received depressing news about the state of
my account, I spotted a man whom I had met once at the
Coryndon Museum. He was considered to be one of the best
wildlife cameramen in the world and his name was Des Bartlett.
This was my first taste of serendipity – the way chance events
can develop in the happiest and most beneficial way – and
something that has been a feature of my life ever since. Des
was cameraman for Armand Denis, a Belgian/American film-
maker who had just been given a contract to make a series of
films for BBC television to be called *On Safari*. A large-jawed,
friendly and restless man, Des remembered me, introduced me
to his new wife Jen and asked me to give him a hand loading
supplies into a huge new Dodge Power Wagon with the legend
ARMAND DENIS CENTRAL AFRICAN EXPEDITION painted
on the doors. I speedily made myself as helpful as possible
and before Des could drive away I grabbed the bull by the
horns.

'With all that work, won't you be wanting some help?'

'Definitely,' said Des, 'and a good mate of mine is coming
over from Australia, as I'll certainly need an assistant.'

'Or two?' I asked hopefully.

'You could be right.' Des grinned. 'Come round to the house
some time and speak to Mr Denis.'

I did and was offered a job looking after the Denises'
expanding collection of wild animals – a cheetah, bushbabies,
meerkats, a honey badger and several antelope. Armand Denis
was a large, jovial man, oozing confidence, who had led an
extremely adventurous and successful life making wildlife
documentaries in many countries around the world. His wife
Michaela, who liked to call herself a white Russian, was a

glamorous creature I had first seen driving a white Cadillac convertible through Nairobi, her signature orange hair flying, with a cheetah sitting calmly beside her. Mr Denis, knowing I had a collection of snakes, asked if I could help him film an egg-eater snake, which, as its name implies, feeds entirely on eggs. It has no teeth, which would only get in the way of swallowing an egg. Its jaws are able to dislocate at the hinges and the two halves of the lower jaw can also spread apart, held together by an elastic ligament. This enables an eighteen-inch, pencil-slim snake to swallow a chicken's egg whole – an extraordinary sight. A couple of inches down its throat several vertebrae have downward-pointing spines, which saw the egg open as it passes. The snake then crushes the shell with vigorous writhing, squeezes out and swallows the contents and regurgitates the neatly folded shell. I told Armand Denis I would provide an egg-eater from my collection and some pigeon eggs. I also suggested I show him and Des my own 8mm film of the action, so that they knew what to expect. This was a deliberate ploy, as my film not only featured an egg-eater feeding but also a boomslang, a species of tree snake, raiding weavers' nests, various other bits of snake action and lots of shots of rhinos chasing the jeep, which had been a source of much hilarity for Nick and me on our trips together. When the film was finished Armand Denis sat back and said, 'What the dickens are you doing just looking after animals when you can take pictures like that? Why aren't you filming them? For me?' Suddenly I felt confident enough to tell him that I had actually done a little 16mm filming, that some of it had even got on to television and that yes, please, I would certainly like to change my job description.

Within days we set off for Ngorongoro and the Serengeti, which at that time were contained in one great national park in Tanganyika, about three hundred miles west of Nairobi. I had made this trip before in my battered old ex-army jeep but now I was Des Bartlett's assistant in a brand-new, open-topped Willy's jeep with that magic CENTRAL AFRICAN EXPEDITION on the side. Jen Bartlett was driving a station wagon and Des a three-ton Dodge truck called the 'Queen Mary', loaded with several months' worth of camping supplies.

Ngorongoro is one of the natural wonders of the world. Over three million years ago it had been a sixteen-thousand-foot volcano, but then its top blew off and much later the cone collapsed in on itself, creating a caldera, a crater two-thousand feet deep and ten miles across. Now extinct, Ngorongoro's rim and slopes are forested with olive and cedar trees heavily festooned with beard moss, while down below is a vast grazing lawn, divided by clear streams, with a large patch of fever-tree forest and a much larger shallow lake. This lawn is mowed by thousands of wildebeest, zebra, eland, gazelle and buffalo, and in 1956 contained over a hundred rhino and sixty lions, along with many smaller carnivores and several large packs of hyenas.

Accommodation was available in log cabins on the rim, but we chose to camp on the crater floor in a tiny one-roomed cabin next to a clear chuckling stream, bordered by beds of watercress. If anywhere deserved that grossly overused name 'Eden' it was Ngorongoro. Rhinos in groups of four or five walked through ankle-deep clover and fields of yellow daisies, the thirty-six lions we knew as the Munge pride sprawled in the sun next to the cool clear stream that gave the pride its name, and five hundred crowned cranes pirouetted and leaped in a mass mating dance on the shores of a flamingo-crowded lake.

We were up at five every day, shivering in the cold air that poured down off the crater rim overnight. Bolting down huge mugs of steaming tea and a handful of biscuits, we were off before dawn. Lions would be either replete and sleeping near a stripped carcass, or, after an unsuccessful night of hunting, would have taken up positions from where they could ambush the herds that would later come down to drink. Gangs of spotted hyenas would be shuffling back to their burrows, their low-hanging heads often covered in bright fresh blood. We noticed that the hyenas were sometimes even bloodier than the lions, but we didn't understand why. Then one day we were able to film several hyenas chasing a wildebeest and, although they failed to bring it down, they had made a serious attempt. Hyenas were meant to be scavengers, not hunters, but these animals were obviously hunting, which could explain the freshly bloodied hyenas we had seen. A couple of years later I filmed a pack of hyenas who proved they were very successful hunters – but this became another of the discoveries I failed to publish. With an Arriflex camera supplied by Armand Denis and no limits placed on my film consumption I was on a very steep learning curve and having a wonderful time.

After a couple of weeks in the crater we moved north-west to the Serengeti. Cresting the crater's rim at around six thousand feet above sea level and driving over the cool, green, rounded hills, you suddenly come over a rise and there spread before and below you is a vast lion-coloured plain that just goes on and on – the Maasai call it Siringet, 'the great open space'. To the east a low range of pale-blue rocky hills dance in the haze, Ol Doinyo Gol, 'the hard mountains'. Tall ashen dust devils move across the land like a gathering of ghosts and suddenly you realise that the other clouds of dust rising from the plain are being kicked up by millions of hooves. The wildebeest are on the move. Your eyes can only take in so much, so you raise your binoculars and, incredibly, you see the herds go on and on until lost in the shimmering heat. It is a scene straight out of the Pleistocene and, five decades later, having spent much of my life there, I still shake my head and become wet-eyed just from knowing that such a place still exists on our crowded planet. There is no sight to match it.

Once down on the plain, driving gets very uncomfortable. The track is a deep trough filled with talcum-fine dust that rolls up and over the bonnet in waves – and in the open jeep I had always to wear a bandana over my mouth and nose. To make things still more uncomfortable, there is a following wind that is impossible to outrun, so you drive in a dense dust cloud of your own making. It is all made worthwhile, though, by the greeting that the Serengeti herds give to vehicles. Whether wildebeest, zebra or gazelle, they will all race alongside, gradually pull ahead and then suddenly swerve across your path and leap wildly over the track just ahead of your vehicle. Out on the open plains, where one side of the road is no different or safer than the other, this is a pointless manoeuvre, which appears to be simply a joyous response to the challenge of a moving vehicle. On that expedition, crossing the great plain took several hours, until a distant mirage slowly metamorphosed into rocky hills and acacia trees, and we arrived at the place the Maasai call Sironet. We camped there near the Seronera river, one of the Serengeti's 'hot spots', an area that – for reasons not yet understood – has a greater density and variety of species year round than anywhere else in the park. Groups of granite boulders are clustered among stands of mature acacia tortilis, the finest of Africa's flat-topped trees, and the tea-coloured river is bordered with palms and stands of the yellow-barked acacia known as fever trees. These trees are almost always found near water, where early explorers naturally tended to camp. When they caught malaria they made a connection with the trees, hence the name, but not to the mosquitoes that bred in the water nearby.

We had only been there a short time when Myles Turner, the Park Warden, suggested I join his assistant Gordon Poolman on a week-long recce to look for evidence of poaching in an area he had yet to explore himself. Gordon was the Kenya-born brother of well-respected professional hunters. Enormously strong, with a .38 revolver in an open holster, to a nineteen-year-old he seemed a kind of iconic figure and I looked up to him in awe. The plan was to follow the south bank of the Mara river down to Lake Victoria, cross the river, then follow it back up on the north side to the Kenyan border, where we would recross it back into the Serengeti. It would all be cross-country, and Gordon and I looked forward to a jeep versus Land Rover competition. We would take two park rangers with us each, but the objective was not to capture poachers, but to assess the extent and size of their operations.

The Mara river flows through superb game country that in the fifties harboured a rhino in almost every sizeable thicket. Buffalo in herds of hundreds thundered off ahead of us and large groups of hippos and crocodiles occupied every quiet stretch of the river. Poaching was firmly established in these ungoverned areas, and there were well-worn tracks leading into the park from the villages outside its boundaries. Across the game trails that led down to the river we found long thorn fences with gaps set with heavy wire snares to strangle any animal passing through. We found some with dead animals in them and two with live zebras held by the neck; one we were able to release, the other so badly damaged that Gordon shot it. In a couple of areas a real hazard for the vehicles were rows of deep pits covered with grass mats and cut branches, many of them surrounded by the bones of animals that had fallen in and been butchered on the spot. On the open plains we found strings of fifty or more home-made rope snares hanging from a taut horizontal line and set for gazelle. The poaching was brazen and on a large scale – these were killing fields. No effort had been made to hide their camps, which were surrounded by masses of meat drying on racks or laid out on the ground. Vultures hunched in the trees and there were even smaller string snares set for them, as their feathers were needed for fletching poisoned arrows. When we arrived at the Mara bridge, Gordon was called back to his HQ and asked me to complete the recce without him. For the next four days the two rangers and I followed the Mara river up through sparsely settled country near the lake, then uninhabited woodlands and on to the wild Lamai section of the park.

Driving along the river as I was nearing the end of the trip, I came across a bizarre sight. Perched high in a great acacia was the rear body of a three-ton Bedford lorry – a huge, low-sided wooden box – with a thatched roof and a ladder propped against it. Following a footpath that led back into the bush, I came upon a few thatched huts and a large man even scruffier than myself, who introduced himself as Ace DuPreez, a South African gold miner. In the ground behind the huts was a deep shaft down which half a dozen Africans disappeared every day and sent up buckets of ore that contained just enough gold for him to afford to pay them to go back down for more. Like many a successful man he had a second home – the lorry body – and, filling a basket with Tusker beers, he suggested we head there for a sundowner. When the talk got around to filming wildlife, Ace told me he had once been involved with a Hollywood crew making an animal capture film called *Zanzabuku*. I remembered seeing the film many years before. Most of it was roping giraffe and rhino from an open jeep that got seriously hammered in the process, but there was one scene that I had never forgotten involving a man who seemed to be determined to die a very messy death.

I started to describe it, hoping Ace could tell me more. 'It was a guy in a tiny boat and some hippos,' I began.

'Ya, nea, that was me!' His reply rocked me back in my dilapidated chair.

I remembered well the disbelief and dread I'd felt as I watched that scene. Hippos have monstrous mouths that open a hundred and fifty degrees and can easily take in a man. They also have razor-sharp canines, thick as your arm and eighteen inches long, and they kill more people in Africa than any other animal, many of them fishermen who venture too close in unstable boats. The sequence that had filled me with such foreboding showed a white man standing up in a very small rowing boat floating slowly towards a school of about forty hippos. As he got closer they began to turn and look towards him, closer still and they started snorting loudly, sending clouds of spray high into the air. All his attention seemed to be fixed on a simple Box Brownie camera he was holding, as he drifted even closer and the hippos began to grunt and bellow, lifting their heads high out of the water with necks arched in aggressive posture. The slow motion, looming inevitability of the scene, was totally mesmerising. Closer still, and now the hippos were porpoising and getting extremely agitated, but the man continued clicking with his Brownie until finally he was in the centre of this swirling snorting mob of highly disturbed two-tonne animals. When the waves they kicked up by their massive bodies swamped the tiny boat, the man fell in and simply swam to shore through the midst of the now frenzied animals. I had never seen such an undisguised death wish.

'That was *you*?' I asked, astonishment mixed with reverence.

'Ya, nea, that was me all right,' he replied.

'But . . . but you must have been out of your gold-digging mind! That was madness; what on earth made you think you could get away in one piece?'

'Agh, Alan, you know, I did it for the money. Five hundred dollars was a lot of money back then.'

'Hey, I know some pretty crazy guys, but no one who would take that risk. You were floating to your death.'

'Yah, nea, that's what they all thought. But I didn't tell them.'

'Hang on, you didn't tell them what?'

'Oh, come on, Alan, you're in the business, I just didn't tell them. Yirra, you know.'

'No, I *don't* know. I can't think what you could possibly *not* tell them to justify what you did.'

'Agh you know these Hollywood types, you know the tricks. They know nothing about Africa so you dream up something that looks dangerous and charge a lot for doing it. They get their film and are happy, and you just don't tell them.'

'Don't tell them *what*?' I almost shouted. 'That *was* dangerous – hell, it was suicidal. What *didn't* you tell them?'

With a grin and the solemn air of a guru imparting a piece of eternal wisdom, Ace confided, 'I didn't tell them . . . that hippos only eat grass!'

I sat there stunned, trying not to giggle in disbelief. Reckless behaviour powered by testosterone, alcohol and the promise of five hundred dollars I could have almost understood. But this? There have been extraordinary stories of small children crawling up to wolves or bears and not being harmed, their undeveloped sense of fear somehow failing to trigger the attack that would have greeted a nervous adult. This simple man, childlike in his belief that animals which 'only eat grass' would not want to hurt him, had seemingly enjoyed such protection, too. Cloaked in innocence and naivety, like a nun walking serenely through an angry mob, he had simply breast-stroked his way through that roiling mass of beasts, climbed the bank and walked away.

All I could manage to say was: 'God, you were lucky.'

'Ya, nea, it was OK, it was lekker,' said Ace. Then, after a quiet, thoughtful moment he added, 'But you know . . . since then I've realised that buffalo also only eat grass – an' man, they blurry dangerous!'

It was a warm night and his stories flowed as freely as the beer, but nothing came close, or has ever come close, to what he 'didn't tell them'. Twelve years later, as I put on my goggles and prepared to be the first man to go underwater with a group of hippos in a crystal-clear spring, I tried reciting the mantra 'They only eat grass', but found it gave me little comfort. And you know what? A hippo bit me.

When I got back to camp at Seronera, Des outlined the order of events for the safari. He and I would spend a few weeks filming, and when Armand thought we had enough footage for a half-hour programme he and Michaela would fly in to a photogenic camp we had prepared for them. The task then would be to get them into safari gear and the open jeep, and make it look as if they had done all the filming. Armand used a wind-up Bell and Howell camera with a three-lens turret for these make-believe shots. He liked it because it had a large silver key on the side, which turned as he 'filmed', giving it all an air of authenticity. Anyone knowing something about photography and wildlife would have realised that hand-holding a six-inch lens you just don't get the rock-steady close-ups for which we had used a twenty-four incher on a heavy tripod. But how many people knew that much back then? Michaela looked gorgeously competent and stood – or more often sat – by her man through all these dramatic moments. We felt sure her amazing orange hair had been created by some Kodak chemist because it looked so wonderful on Kodachrome, which was very biased towards the red end of the spectrum. The end result was, for its time, great television, and *On Safari* went from strength to strength.

Des and I spent several more weeks working out of Seronera, getting the first film of a leopard taking its kill – a gazelle – up into a tree, baboons hunting baby gazelle, a zebra giving birth and much more great footage. Then shooting finished and Des moved up to Uganda, while I accompanied Armand and Michaela Denis on a trip to northern Kenya. We stopped off en route near Isiolo for tea with the legendary Game Warden George Adamson and his wife Joy. By 1957 Joy was already renowned for her several marriages, her paintings of Kenyan flora and an irreplaceable collection of her portraits of Kenya's many tribes in traditional dress. Now she was rearing a lion cub called Elsa; she had already been on television with her pet and as a result was on the cusp of major celebrity. Michaela's glamour and international fame obviously stirred envy in Joy's sun-scorched little soul and this led to an incredible battle of the divas. Armand, George and I sat sipping our tea and shuffling with embarrassment as the two women went at it – all in syrupy, civilised tones, yet dripping vitriol.

Joy to serve: 'I hear that your films are doing very well, Armand' – pause – 'but that people do not like Michaela.'

'Oh, I don't think that's tr . . .' Armand started to say.

'People say she looks false and out of place in the bush.'

'Well, I get eighty per cent of the fan mail,' Michaela returned, 'and obviously people like to see a good-looking woman doing adventurous things.'

'Ah, but those people are beginning to realise that you just go out and pretend to do these things – that other people do the filming – isn't that true?'

'I'll have you know we have just spent two weeks on the Nile,' volleyed Michaela.

'Oh yes, I hear there is a fancy new hotel at Paraa – what is it like?'

We men in the room were shuffling and cringeing, and George quietly suggested we might like to look for tracks around the waterhole, so we slunk out, leaving the women to their battle, and in doing so obviously removed any need for civility. The decibel level rose and before we had reached the waterhole there were loud explosions, roof tiles and shrapnel came flying overhead, and thick black smoke poured from every broken window. We did male-bonding things like kicking at elephant dung for a while until Joy appeared on the veranda and called out sweetly, 'Won't you have some more tea?' That was my first taste of the power of this new medium of TV to create instant celebrity – and of what celebrity can do to a once normal being.

My next assignment for the Denises was in the western section of Tsavo National Park where the Warden reported there was a lot of action around Kilaguni waterhole. This was a tiny natural hole on the edge of a lava flow shadowed by a huge fig tree. A year after I filmed there a lodge was built nearby and an artificial waterhole created, but the elephant and rhino preferred the water in the natural hole so it was filled in and they had no choice but to drink old bathwater in front of the lodge. That, I believe, is called progress.

I built a small platform high in the tree and the next morning I was there at five o'clock. In the dark I climbed as quietly as I could, hauled up, then set up my tripod and camera, and laid out my lenses – ready for anything. Anything except what I got. As the sky began to lighten I became aware that I was not alone in the tree. Above me a large troop of baboons were waking up and dawn is the time for the first pee of the day. That alone would have been pretty horrific, but when they saw me blocking the way down they felt trapped. Now they weren't just having little morning piddles, they were pissing themselves with fright and soon the rain turned to mushy hail as fear loosened their bowels. I tore off my shirt to cover my camera but the artillery had got my range. Some of the bigger males

came down as close to me as they dared, then flung themselves out to land in a thorny bush below. Feeling they were being abandoned, the females and young started to shriek in panic, voiding everything they had left and tumbling out of the tree like so much rotten fruit. I spent every day for the next week up on that stinking platform, but I started later and made sure every morning that the baboons came down before I went up. Fortunately I was staying in a friend's camp where there was plenty of hot water, and I had industrial-strength showers every night. But there was obviously a lingering fragrance about me, for when I got back to Nairobi Nick greeted me with: 'Long safari, heh, and you've been running around with that baboon again!'

I worked for a year with Des and Jen in various national parks around East Africa, then Armand Denis moved his operations to Southern Rhodesia and I decided to stay in Kenya. I managed to pick up a couple of jobs guiding photographic safaris, but then one day received a telegram from the Serengeti: 'A plane has been chartered from Caspair for you. Please come to Seronera to discuss filming work. Dr Bernhard Grzimek.'

Wow! I had heard of the father and son Grzimek and their work, and was thrilled to have the chance to meet them. I flew down the next day. Grzimek the elder was director of the Frankfurt Zoo and had been asked by the director of the Tanganyika National Parks to help map the migration route of the great wildebeest herds on the Serengeti. New park borders were about to be drawn up and it was essential to know the herds' route. At that time some quarter of a million wildebeest seemed simply to disappear from the plains during the migration and no one knew where they went.

Grzimek and his son Michael had bought a Dornier aircraft, painted it in zebra stripes, learned to fly and, with the ink still wet on their pilot licences, had flown it out to the Serengeti and started following the herds. Alongside their research they also wanted to make a film and had brought out a documentary cameraman from Germany. He was a first-rate technician, but knew nothing about wildlife. He would climb on the roof of the Land Rover to set up his tripod, scaring off animals that were unused to seeing people or cars. After a couple of months with little success he was sent home and Myles Turner, the Warden who had sent me on that poaching recce up the Mara, suggested me. My interview with Dr Grzimek was short and to the point. After the ineffective trial run with the German cameraman they were months behind schedule and urgently needed a replacement. Myles had given me his endorsement and that was good enough for Dr Grzimek. Did I want the job? Oh yes, I did – so much so that I hugely understated my worth – naming a sum that was swiftly accepted by Grzimek. When I turned up for work three days later with my jeep and sleeping bag, I learned that Myles had been speaking up for me. The Grzimeks were qualified zoologists, he had argued, but I knew much more than they did about the wildlife and geography of the Serengeti. I was not just a cameraman, I would, in effect, be their guide and as such should be paid guide rates, so I was quietly told by Dr Grzimek that my salary would be twice what I had requested. Good news indeed. It was only then that I learned that this was not just a TV documentary, but a full-length 35mm film for the big screen. Could I handle it?

3

The Deep End

During filming I was to share a large aluminium Uniport hut with Dr Grzimek, his son Michael and Herman Gimbel, a school friend of Michael's who had joined the project as an assistant cameraman. Michael was a tall, good-looking lad a year or two older than me who enjoyed life mightily and we hit it off immediately. He was a natural pilot and flew the zebra-striped plane with great panache. Its German registration was D-ENTE, meaning duck – good advice to those on the ground when Michael was flying. Initially, Dr Grzimek appeared to be a serious character, oozing gravitas, very proper and formal. Many years later I overheard a conversation between him and Gordon Harvey, a senior Park Warden, that showed only a slight relaxation. Grzimek had just been made a professor and Gordon asked him, 'So how should we address you now, Doctor or Professor?'

'How long have we known each other?'

'About six years.'

'And how old are you?'

'Oh, sixtyish – much the same as you.'

'Ye-es – I think you can now call me Bernhard.'

Once we'd had some adventures together and got to know each other he became less Germanic and revealed a good sense of humour.

With the exception of lions, which hunters and early

film-makers had been approaching in cars for years, often throwing a carcass out of the back to feed them, the other Serengeti species were shy and difficult to approach. The 35mm equipment was bulky and heavy, so I strapped a tripod over the jeep's passenger seat, which would allow me to film without moving from behind the wheel, giving great flexibility and a fast set-up. One of the lenses I used was a five-foot-long thousand-millimetre monster that stuck out over the jeep's hood like a bazooka. From a distance the rig looked as though it belonged on the battlefield rather than in the bush. It waved around in the slightest wind and jumped along with my heart-beat if I pressed too tightly to the eyepiece.

I was given a free hand to film as many wildlife sequences as I could, and would spend every day locating some promising situation and then slowly working my way close enough for filming. The animals were far shyer then than they are today – now that several generations have grown up with tourist vehicles as part of their world – and collecting good footage was a slow, arduous process. I filmed antelopes fighting, giraffe mating, wild dogs hunting, then returning to their den where they would call out their pups and regurgitate meat for them, lions hunting or just quietly going about their family business. In addition to the wildlife filming we were covering the Grzimeks' research work, ranging from aerial attempts to count the migrating herds and map their movements; airborne anti-poaching operations and Michael's off-field landings to collect grass and soil samples from all over the park to see if the animals' movements could be tied to vegetation changes. We were also experimenting with using tranquillising dart guns. It made for good cinema, but our dart-gun technology was very primitive and the drug we were using, nicotine salycilate, extremely dangerous. We found the drug worked on wildebeest, but then it took many hours for the effects to wear off, and we would have to stand guard against predators half the night as the poor animal walked round and round in endless circles until it recovered. We managed to put large yellow collars on several wildebeest, but it all took so long and was so risky that when we moved on to zebra we decided to try to rope them in animal-catcher fashion instead. This involved a man in the back of an open vehicle wielding a long bamboo pole with an open noose at the end. The pole was held pointing over the front of the car until it was alongside the animal, when the noose was swung back and over its head.

Michael was nominated the catcher and we raced off after a zebra, the jeep camera car driving right alongside. Michael had the pole pointing forward at full extension when the car went over a bump and the pole dipped down and dug into the ground. The butt end caught Michael in the throat and he was thrown from the pickup, effectively pole-vaulting on his Adam's apple at twenty miles an hour. He was very lucky, and badly shaken by such a close shave, but apart from the loss of a lot of blood, he was unharmed. We decided to change our technique to one put forward by my old idol, Gordon Poolman, who suggested we simply reach out and grab the zebra by the tail from the back of the jeep, something he had done in his days catching animals for zoos Once the animal was brought to a halt, more people could hang on to the tail while another grabbed an ear and the bottom jaw where it narrows behind the incisors. We would then slip on one of the colourful nylon collars that were intended as markers so we could track the animal's movements from the air, but in the vastness of the Serengeti, with only ten animals collared out of a quarter of a million, this never happened. The only collar we saw again was on a wildebeest being eaten by lions. With that bright yellow splash of colour he must have looked to them like steak with a dash of mustard.

One day a little later we were flying along the Rift Valley to the east of the Serengeti, filming some aerial sequences over

Lake Natron, when we spotted thousands of flamingos nesting on the soda surface. East Africa is close to the equator but the high temperatures on the coast drop as you move inland and climb up to the plateaux at five thousand feet above sea level. Then, when the land suddenly drops to the floor of the Rift Valley at around a thousand feet above sea level, the temperatures soar back up. Consequently evaporation on Lake Natron is high and with no outlet the water has become increasingly alkaline over millennia. The solution of sodium sesquicarbonate is so strong that it crystallises on the surface to form a crust that in places is metres thick. From above it looks like the surface of an alien planet, a range of shimmering pinks and purples, and sinister black pools.

The flamingos had scraped up little mounds of soda about a foot high before placing their single egg on top, and there were tens of thousands of them sitting there. This spectacular sight had only ever been seen before by the ornithologist Leslie Brown, also from the air. As no one had ever visited a colony on foot, let alone filmed the activity, we made plans to get out on to the soda flats and record it. Michael, Bernhard and I then flew in and landed on the soda crust as far out from the shore as seemed safe.

The temperature out on the flats can reach 165°F (74°C) and the glaring white surface reflects the light so you get a double dose of sunshine (which can seriously sunburn the underside of your chin). Carrying the heavy equipment would have been exhausting in that heat, so I had made a sledge from an old aluminium fuel tank, and we persuaded a Maasai herdsman to bring along a donkey to haul it. It went well for about half an hour, but the donkey soon became more and more spooked by the mirages that shimmered over the surface of the lake. Any small piece of driftwood or protuberance on the surface loomed up like a shadowy figure dancing in the heatwaves. The last straw for the donkey was a line of rhino tracks. What that animal had been doing this far out on the soda I cannot imagine, but the crust pushed up by its heavy tread produced a phantasmagorical effect of a long row of pulsing fence posts sticking up out of a sheet of tremulous water. This was too much for our donkey and it took off at high speed for the shore, capsizing the sled and scattering our valuable gear over the highly caustic crust. The Maasai took off at even higher speed to try to catch his donkey which, thinking it was being relentlessly pursued by the sled – as, indeed, it was – looked like it would keep going for ever.

We gathered up the equipment and realised that our canvas filming hide had gone off with the donkey so we couldn't film even if we had been able to get close to the colony. As we headed back, the sun beat down and bounced up at us from the crust, and the acrid dust clogged our nostrils and scratched our throats raw. We had planned to spend the night next to the plane, and there we found our Maasai with his donkey, the sled and the rest of our equipment. After rewarding him and cleaning all our gear, we sat on our sleeping bags around a small driftwood fire having a beer and cooking sausages. It was a sublime evening, the sun going down with a spectacular flourish of colour behind the ten-thousand-foot volcanic cone of Ol Doinyo Lengai – the Maasai's Mountain of God.

For once we had nothing to do, so we talked about how the research into the migration routes was progressing, and how a pattern was emerging to the wildebeest's movements: the calving out on the plains early in the year, then the move north-west in May when the plains dried out, and then a few months spent further north before heading back to the plains in November. We had only seen one full circuit, so did not know if that had been typical, or if it varied from year to year. We went over what sequences we had in the can and what we still needed for the film. As the Grzimeks would have to return to Germany at some stage to view what we had shot so far, I would carry on alone. Gazing into the fire, we talked late into the night of future plans for a TV series that I would shoot, and for a research station on the Serengeti, until a sudden crash of thunder made us look skywards. A huge storm had stolen up on us while we were weaving dreams and the first heavy drops started to fall with loud smacks on the plane. The wings were large enough to give us shelter while we piled the gear inside and there was plenty of room for us to crowd in there too. But I was worried about the effect of rain on soda flats, which I'd seen before. They are so level that they flood swiftly and it takes very little rain to turn them into an impossibly slippery and sticky quagmire – and this wasn't just a little rain, it had turned into the mother of all storms. Over the loud hammering on the fuselage and wings I shouted to Michael that we had to get the plane higher up the shoreline and on to the grass or we could be stuck for days. He jumped in and started her up, put on the landing lights and, with Bernard and me pushing on the legs, he gave her full throttle. The surface was already deliquescing and we slipped and fell repeatedly, losing our footing and being knocked down by the powerful propeller blast. Laughing and shouting, we inched the Duck forward until finally she climbed out of the mire and on to firm ground. Bernhard and I were covered in corrosive soda mud that burned skin and eyes, but we stripped off and were washed clean by the unrelenting rain. We had failed to get any footage of nesting flamingos, but it had been a peerless day, full of extremes, and a great bonding experience. Sadly it was also our last time together.

A few days later I was back in Seronera, waiting for Michael to come and collect Herman and me. We expected him to fly in that evening, but he never arrived. This wasn't unusual and as we had no means of communication until the early morning radio call there was no point in worrying. Before light the next morning, however, I woke to agitated male voices and the sound of women sobbing outside the house. I leaped out of bed just as our host Gordon came in and said in a breaking voice, 'Michael has crashed and is dead!' It was the first time I had lost someone close to me and the reality took a while to sink into my numbed mind. Michael? That dashing pilot? Dead? Somebody must have got it wrong.

Michael had left his father in the cabin in Ngorongoro the afternoon before and had decided to do some aerial filming on his way to collect us. He had been flying down the Sanjan Gorge, a spectacular cleft in the Gol Mountains where hundreds of Ruppel's griffon vultures nest on the cliffs, when one of those twenty-pound birds had slammed into the wing. The control cables for the ailerons, the flaps that raise or lower the wing, run just inside the leading edge of the wing and the vulture had made such a deep dent that it jammed them. This had caused the plane to roll on to its back and dive into the ground, where it smashed to pieces, scattering across the grass where the Sanjan flowed out of the gorge. Michael was killed instantly. A man who had been drilling for water in the area brought Michael's broken body up to the Ngorongoro headquarters, from where a note was sent down to his father, who was alone on the crater floor. With the National Park's approval our small group buried Michael the next day on the rim of the crater, with a wonderful view out over the bowl and the highlands beyond. His headstone read: 'He gave every-thing he had, including his life, for the wild animals of Africa.'

Michael's death was a terrible blow to everyone on the Serengeti. Shattered as he was, Bernhard Grzimek was determined that we should finish our film as a memorial to his beloved son. Within a couple of days he returned to Frankfurt, where he would have the heartbreaking task of looking at all the footage and would let me know what was needed to complete the film. I was to carry on alone, thrown in at the deep end. For the next five months I worked flat out. There were aerial shots that we had not had time to shoot, anti-poaching operations and Maasai ceremonies still to record. We wanted an animal birth and a rhino hitting the car among other things, and Ogiek hunters following the honey guide bird and collecting honey and boiling up acocanthera bark and twigs for their deadly arrow poison.

Also on the list was one sequence that I kept deferring. Grzimek was a canny showman and he wanted some scenes of naked women in the film, which he knew would enable him to get his conservation message across to a larger audience. Back then, screen nudity was banned, tolerated only if shown in an anthropological context. What Grzimek wanted was shots of Maasai women bathing naked in a river pool, then rushing out of the water when they realised a leopard was watching them. I was unhappy about this. It wasn't that I was prudish, I was as keen as the next twenty-year-old to see naked women, but I had a good relationship with the Maasai and didn't want to ask them to do something that I could not satisfactorily explain. I was hoping Grzimek would drop the idea, but when I had ticked off everything except that one sequence he said if I wouldn't shoot it he would find someone else who would. Then it dawned on me that once naked, those girls could be any tribe under the equatorial sun, which meant I didn't have to compromise my friendship with the Maasai or gain a reputation with their modest womenfolk as some kind of voyeur.

Back in Nairobi I had a contractor doing odd jobs for me and I roped him in for the task: gather a group of buxom wenches prepared to splash around naked in the Mbagathi river fifteen miles outside Nairobi. Have them at that secluded spot at a certain time and an hour later I would cross all their palms with silver. These were colonial times with colonial social mores. I'd chosen what was then a remote location because what I was about to do *simply was not done by a pukka chap*. Man, what uptight times those were! Anyway, the girls turned up on time, splashed and cavorted in the sunlit acacia-fringed pool and on cue they fled, up and away from the phantom leopard. When the time came to pay them they couldn't believe that the show was over. Apparently they were all ladies of the night and thought I had a more involved agenda – I think they saw the swimming bit as a sort of 'Have that one washed and sent to my tent' scenario. Finally my list was complete and Dr Grzimek was happy. The resulting film, *Serengeti Shall Not Die*, is widely considered to be one of the most influential wildlife films ever made and six months later won an Oscar for Best Documentary at the 1960 Academy Awards.

Now that I had time to catch my breath the loss of Michael sank in properly. It was a bitter blow, from which I found it hard to recover and, worse still, it was a harbinger of other deaths, for over the next year or two I was to lose many more good friends, several of them to aircraft crashes. Hardest of all was losing my childhood friend and soulmate Nick. He had borrowed my jeep for a trip up to his forest station and rolled it on a slippery road. In great pain, Nick drove himself to the Nakuru hospital but no one noticed that his spleen had been ruptured and he died later that night. Losing him was numbingly painful and I wandered the bush for days around Voi, where we had spent our school holidays birding with old Myles North, trying to come to terms with the loss. Instead of all the major climbs and adventures we'd shared, one small incident came repeatedly to mind. We had been camping rough behind the Ngong hills, and sitting quietly eating an early dinner when a tropical boubou shrike called out close by. Boubous live in closely bonded pairs and keep in touch in the thick bush with antiphonal calling sessions. The male gives a 'clockclock' immediately answered with a 'wheee' from his mate – their calls so closely synchronised that it sounds like a single bird. This day there was a 'clockclock' but there came no reply. A long pause – maybe she had her beak full? Another querulous 'clockclock?' and again no reply. Yet another unanswered 'clockclock', then two more in quick succession. There followed a prolonged silence, she *must* have her beak full, then finally another really urgent 'CLOCKCLOCK' followed instantly by an exuberant 'wheee!'; three further rapid celebratory duets, then silence. There was a great exhalation of long-held breath and laughter from both Nick and me. We had sat there quietly, listening and understanding. We'd both felt the male bird's initial anxiety, his not wanting to sound too concerned, then his desperation to get a reply and finally his delight at being answered. All ridiculously human constructs to be sure, but it was the shared understanding and appreciation of that small event that made me realise how close Nick and I had been, and how much – and for how long – I was going to miss him.

In the midst of all this grief I was moving towards another friendship that would eventually lead to the same kind of unspoken and shared understanding of nature. I had first seen this young woman at the annual Nairobi Agricultural Show, and she had stunned me with her blonde beauty and long-legged grace. When I asked those around me who she was I was told not to get excited. She was Joan Thorpe, Kenya-born, and her father would consider me far too raggedy-assed to be allowed anywhere near his daughter.

The next time I saw her was several months later at the Ngorongoro crater. Her father was a successful coffee farmer and also ran one of Kenya's first photographic safari businesses. Joan had been to a Swiss finishing school, but you can't take the country out of the girl, so soon after her return she had joined her father's safari company. On this particular day she had come skidding up to the log cabin lodge in a large safari wagon with a cage full of chickens on the roof. Most of the camps and lodges around East Africa in those days were rather primitive and DIY, so Joan did all the cooking and the chickens were her meals on wheels. Tall and slender, she started hauling heavy American-sized luggage out of the truck and assigning it to various cabins. Running over to help would have been a bit obvious, so I pinned my hopes on dinner that night as I would be eating at the lodge with the Warden and his wife. As luck would have it they knew Joan's father well, so he and his party joined us. Over the course of the meal I managed to say that I had been camping down in the crater for two months, so could tell them where to find lion cubs, a big bull elephant, a wild-dog den etc. etc. – boy, did I go on. Joan's father, known as 'Thorpo', said that as the next day was a rest day for the clients, why didn't I take Joan into the crater and show her everything, so that she could guide the safari when they went down the following day? I thought that was a brilliant idea.

As a treat that evening the Warden's wife gave me a packet of butter, a great luxury, but in my dizzy state I put it in the glovebox of my jeep and forgot about it. The next day, as I drove down the winding crater road with Joan, the butter melted and dripped down her legs. She didn't mention it, but many moons afterwards, when we knew each other better, she said that she had thought it was rather a literal way of buttering up a girl.

Two days later Joan's safari moved on and we had no further contact for several months. On my return to the capital I bumped into Thorpo, as one did in those days, at the Thorn Tree, the outdoor coffee shop and restaurant that was Nairobi's meeting place and the world's crossroad. He told me that Joan was having a hard time trying to raise a baby elephant they had pulled from a mud hole in Northern Kenya. He knew I had experience raising wild animals so maybe I could help. Why didn't I come out to his house and have a look?

Back then it wasn't understood that very young elephants are unable to digest the fat globules in cow's milk or milk powder mixtures. The secret of the correct mixture for baby elephants had yet to be discovered by Daphne Sheldrick, the amazing lady who has since bottle-raised well over a hundred elephant orphans. Joan and I took turns feeding and giving the animal the constant companionship that it needed but it was no use. For a week it took its bottle well and drank plenty, but it was not getting enough nourishment from it, and the little fellow slowly faded away and died. A baby elephant is such a magical creature, playful and affectionate, covered in soft hair, with tiny pink toenails and a floppy, uncoordinated little trunk, which constantly seeks reassurance. To watch one slowly deteriorate and die is a shattering experience that deeply affected both of us. Greatly saddened and exhausted, we went out for a meal together – the first proper food we had eaten for the whole time. We had some wine to drown our sorrows, talked of our shared pain and agreed that we would like to see more of each other in happier times. Here was a girl who loved camping out, bathing in rivers or rock pools, was more interested than concerned when elephants investigated her tent or when she was trapped in the canvas shower by inquisitive lions. I was in love, and those happier times would come.

Guiding month-long photographic safaris with her father took Joan to the national parks in all the East African states, and beyond to the Congo and Zanzibar. I was moving around too, but we somehow managed to meet whenever her trips took her to areas where I was filming. She enjoyed her long safaris, but she longed to live the way I did, camping for months in one place and really getting to know the country, rather than hurrying on to the next destination. Over the following year we saw each other for just a couple of days at a time when our itineraries overlapped, all heavily chaperoned by her father. It was a fragmented courtship, but in those brief times together we had no doubts about where we were headed. We wanted to get married, and felt the news that *Serengeti Shall Not Die* had earned an Oscar had given me enough standing to ask Thorpo for his approval. Our dream was to get married in the Serengeti, but we could not find a preacher with jurisdiction there, so settled for the Nairobi cathedral.

Joan had a lovely pale-sage dress she'd had made for our planned bush wedding; but she didn't feel the need to change to white for the cathedral. Des Bartlett was my best man and we drove away from the reception at the Thorpes' farmhouse to reveal a steaming pile of elephant dung courtesy of one of my less civilised mates.

In a heavily laden Land Rover with a large trailer, we were off on our honeymoon to start filming for some special programmes on conservation problems the BBC had commissioned from Armand Denis. The plan was to stop for the night at Hunter's Lodge, a pleasant spot on the edge of a chain of clear, spring-fed pools. As it was normally half empty I hadn't bothered to book. That night it was full, of course, so we bashed on in the dark to Mac's Inn, a rather scruffy roadside establishment run by 'Mac' MacArthur, an ex-game warden of considerable repute, but who obviously wished he had never got into the hospitality business. This was a bad start to a fairly disastrous first week in which I discovered that the huge, ex-army 'honeymoon' tent that I had bought must normally have required a platoon of troops to raise it, for even with the help of the Land Rover's winch it still took the two of us all afternoon. Then, on our second night, Joan was stung by a scorpion, one of those little yellow jobs that have an excruciating sting, and I saw a new and unexpected side of her nature. I had once been stung by a similar species and had whimpered with pain for several hours. Joan just said 'Oh', took a couple of aspirins and went quietly to bed. This was a theme with her. Whether she had just fallen face-down into a giant nettle bed, had a four-inch thorn come out of the top of her foot, or rolled our Land Rover on to its roof, I would never hear anything more dramatic than a quiet 'Oh'.

We camped in the filigree shade of spreading acacias on the bank of the Tiva river in Tsavo East National Park. The Tiva is a wide sand river that runs through thick bush country dotted with great baobab trees grasping at the sky with their short thick arms. It visibly flows only after heavy rain upstream, when it often comes down in a six-foot wall of muddy red water carrying the bodies of animals caught up in the flood. The rest of the year it is a hundred-yard-wide stretch of burning sand, but below the surface the river continues to flow and in certain spots the elephants come to dig. The area was famous for its huge tuskers, bulls with ivory weighing well over a hundred pounds a side, which would appear out of the shimmering heat of the bush, their measured tread speeding up as they came out on to the river bed to where they knew there would be water. They would stick their long thick tusks into the sand as they patiently dug down with their trunks to the water three feet below.

I filmed all this from a hide, a four-by-four canvas cube with small windows through which to film. There was no cover for the hide, it sat out there on the sand, camouflaged with branches, but still a totally alien and geometric intrusion among the random natural shapes of driftwood and sandbank; but the elephants paid it no heed. Drinking was a painstakingly slow business. The animal would push its trunk as deep as possible into the hole and suck it full, then he'd withdraw his trunk, hold the tip closed and hang it down full length to allow the sand to settle at the bottom. After a while he would open the tip briefly and give a little flick to clear the sediment, then pour the remaining clean water down its throat. Quenching their thirst with the forty or so gallons that these tuskers required sometimes took two full hours, one trunk-full at a time. Often after drinking one of them would walk curiously towards the hide, extending an investigative trunk, then shake its head violently, shedding a cloud of red dust and making a loud, hollow clapping with its ears. Then it would move off, totally silent on the soft sand, leaving me and the clearly audible thudding of my heart.

Soon the elephants would be joined by the other species that had been waiting in the acacia shade on the riverbank. First were the rhinos, their sides stained black with the blood-filled droppings of the biting flies that flourish when the animal is in poor condition, as these certainly were. In their haste to drink their front horn often caved in the sand and filled the hole so that they had to dig with their forefeet like dogs, sometimes squealing with thirst and frustration. Then came the herds of buffalo, moving slowly, exhausted, their ribs and hip bones jutting out of their loose hides, dragging their hooves as they queued up around the few available holes. There was nothing between myself and all these creatures but ten yards and a piece of canvas, but I never once felt threatened. In fact, I felt accepted and very much a part of a scene I understood better with every cramped and sweaty hour I spent observing it.

*

Whether watching over a nest from a platform high in a tree, or concealed in the bush overlooking a waterhole, working from a hide is a totally different experience from that in a vehicle; it's where I've had some of my most moving, successful – and frightening – moments. In a hide, with the very limited field of view through the small, forward-facing windows, you rely on your other senses to keep you aware of what is happening out of sight. We humans have lost most of our sense of smell and can only imagine the rich olfactory world experienced by a lion or a rat, but in a dark hide on a riverbank these odours still tell their story. Your nose comes alive and the rich smell of ripe figs and baboon dung is mingled with the pleasant farmyard waft of buffalo and the acrid trace of a nervous mongoose. The cinnamon scent of drying fig leaves and the sweet fragrance of flowering gardenia or acacias are swept away by the violent assault of a lion defecating a hundred yards upwind. Your ears are also now attuned for the smallest clues as to what is going on outside. Once a large tortoise, breathing heavily, made so much noise that I was convinced it was something much bigger slowly creeping up on me. The distant snort of an impala that has seen a predator; the change in pitch of the wildebeests' grunting that tells you they have relaxed, and there are a couple of zebra barking who will probably lead them down to water, so make sure the camera is ready. The calming calls of doves hour after hour lull you into drowsiness, only to be snapped wide awake by the hissing call of oxpeckers. Flocks of these birds ride on the backs of buffalo and rhino, and the alarm calls they give when they see something suspicious alert their hosts, so when you hear that sound nearby in thick bush you look for the nearest tree. The hide wouldn't be molested by buffalo or rhino, but in order to avoid scaring them and have them thundering off, you stay very still and patiently delay getting out for that much-needed pee.

In a hide on the sandy bed of the Garamba river in the Congo, I have filmed – all within twenty yards – the exquisite Egyptian plover hiding its chicks by burying them in the sand right in front of me; behind them a goose with six goslings paddling among a school of snorting hippo, the little fluff-ball goslings surfing on the hippos' bow waves; behind the pool, five hundred pairs of spectacular bee-eaters diving in and out of their nest holes in the sandy bank in a swirling, screeching mass of carmine and turquoise. Alone and unseen in the midst of such glory I become an element of that scene and feel immensely privileged.

On another memorable occasion I was in a hide nestled among the roots of a giant fig tree on the Grumeti river in the Serengeti. Floodwater had eroded the bank, exposing a lattice of thick roots where some small rodent had been burrowing. I had been waiting for hours for the wildebeest to come down to drink at a pool that was the home of a sixteen-foot croc. Since the hide was in deep shade, I had chosen one with an open top – unsuitable out under the sun but ideal for this cooler spot, allowing me to spend the fruitless waiting hours watching birds eating figs in the tree above. I was lying back on the roots, looking up at a barbet and listening to the various scratchings and rustlings in the leaves around me, when I felt a movement by my foot. It was a gentle nudge, probably a small foraging monitor lizard – I was too comfortable to move and look down. Then something was moving across my ankle, something heavy and sliding. It could only be a snake. Quite a big snake. I was so familiar with snakes that I felt no great concern and guessed it was almost certainly a python of which there were many along the river and, whatever it was, snakes don't just bite a recumbent human. That human would have to make a startling move to provoke attack and I wasn't about to do that. It was crossing over my ankles now and slowly moving up the side of my calf. Time to check it out. I raised my head very cautiously and awkwardly to take a look. It was a black mamba, as thick as my wrist and who knows how long – I couldn't tell, as most of it was still outside the hide.

Mambas are the most beautifully designed killing machines.

Even when feeling seriously threatened one can only admire their superb symmetry and fitness for their role. They have a large mouth that can open wide to bite and big, hollow fangs at the front of the jaw so that even a glancing bite will deliver the injection of poison that can kill a mouse in seconds and a man in hours. Their long, flat-sided head is best described as coffin shaped, words that easily spring to mind when one is nuzzling your leg. Clamping my jaw tight, I lowered my head so that I was more comfortable but could still see him. I knew he would not bite me unless I made a hasty move and I just had to keep believing that. He was now pushing his long head under the exposed roots, searching for lizards or rodents, the forked tongue with which he picked up scents flickering constantly. Moving methodically from side to side he would check under the roots, then push his nose under my leg and search there. He was slowly working his way up the bank and soon he was burrowing under my thigh. Then my waist. I needed to keep him in sight, but holding my head up was extremely uncomfortable and I was sure my neck muscles would soon cramp or start to tremble. Every now and then he would find a hole among the roots and slide in for a few inches, tempting me to move while his head was out of sight, but each time, finding nothing, he withdrew quickly and continued his search.

Until now my fear had been controllable, it had even been fascinatingly instructive to observe both his behaviour and my own, but as his questing head moved up the side of my chest towards my armpit I was having serious doubts that I could keep my cool. If he started searching around my head or face I felt I would lose it. I thought of him quickly swinging that long head to look curiously at my widely staring eyes. Snakes are unable to blink. Would I be able to outstare him? Maybe I should shut my eyes tight now, so the question wouldn't arise? Knowing the price I might pay if I moved, would I be able to stay still while he checked out my face?

He was at my shoulder now. I was sweating and his head was so close that I had to look through the reading section of my steamed-up bifocals. I could see most of him now, stretching down from my shoulder and round my feet, with his tail-end still outside the hide. He must have been about eight feet long. Then he moved a little to the side and out of the corner of my eye I could see he had found an interesting hole. He slid slowly in for a foot or so, paused, then surged deeper for at least another foot, as if he had found some prey. If so, with luck he would even have his mouth full. It was now or never. I had to move before he came out. If I had been in one of my usual hides it would have been impossible to scramble out of the zipped-up side entrance in a hurry, but because I had chosen the open-topped model that day I could get away provided I could make a vertical take-off from a reclining position and clear the four-foot canvas wall. It looked impossible, but because the hide was on the steep riverbank, the front wall was lower down the slope, giving me a small advantage, and it is incredible what a body can do when fuelled with industrial quantities of adrenalin. Feeling positively Olympian, I hurled myself at the sky. Quite a lot of me cleared the top of the wall, but my thighs did not, so I tumbled down the bank head first into the fetid water below. The next thing I knew I was across a sandbank and on the other side of the stream. As I flew I heard a great thrashing in the dry leaves as that wonderful snake accelerated away.

During our long honeymoon on the Tiva river Joan and I spent the nights around the full moon watching over the Makoka waterholes in the river bed from behind a low wall of rocks. There was no break in the pageant of elephant, rhino and buffalo coming down to drink all through the night. Different herds of elephants greeted each other with loud rumbling roars and trumpets, then got down to the serious business of drinking. The gleam of big ivory in the moonlight and the grunts and

Sailing over Kilimanjaro

Parked out on the plains, my Cessna was the best vantage point for this cheetah

A small two-seater 'air-chair' balloon that I used for some filming

Registration 5YOYO in recognition of my early flying techniques

Looking down on Kilimanjaro's crater

Chilling out in the amphicar

Joan with Wamba, her baby elephant

Joan with some of our bongos
in quarantine in Naivasha

The first bongo I reared
successfully

Minnihaha the striped hyena and
a young Sally the hippo

I didn't wait around to see how the relationship developed...

Collecting elephant hair to make
Joan a bracelet

A fly-swat was all I had to hand.
All these shots were taken
moments before I had to run

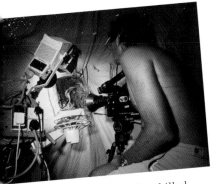

Filming inside the yellow-billed hornbill nest

A very unstable camera vehicle

Filming a 20ft termite tower at Lake Baringo in Kenya

Filming wildebeest on the move

The 1000mm lens on a 35mm camera used to film many scenes in *Serengeti Shall Not Die*

One of the thousands
of flamingo chicks
we were able to save

'So that's where they
come from.' Joan shows
flamingo eggs to a Maasai

Breeding is over,
leaving hundreds
of thousands
of empty nests

Me (bottom right) filming a side vent
of Mount Nyiragongo volcano
in Zaire as it blows its top

Filming the spitting-cobra sequence

When the migrating wildebeest herds spent three days crossing Lake Lagaja in the
Serengeti, some 2,000 calves either drowned or lost their mothers in the pandemonium

A mother elephant tries to raise her dying calf during a severe drought in Kenya's Tsavo National Park

When Lake Katavi in Tanzania dried out tens of thousands of 3ft catfish were crowded into a single pool of rapidly drying mud

Elephants had been tusking this baobab for some time for the moisture in its pulpy wood. The weakened tree then fell and crushed this giant

sighs of contentment from the great shadowy figures as they drank the cool water was magical.

Just a couple of weeks into our marriage, Joan was enjoying her new life and happy to be living in a tent. She had spent time in the bush on her father's safaris, but never in such rough conditions and risky situations. At one point a lion roared on the other side of the river and I recorded the powerful sound on our new tape recorder. To make sure I had a good recording I played it back through the headphones, but they were not fully plugged in, so the sound emerged from the speaker instead and sent an answering roar out into the night. Hearing a challenge coming from our little rock refuge, the lion raced across the moonlit river bed towards us, a swift dark shadow on the silvery sand, grunting angrily. I stood up as tall as I could, shouted and waved my arms, and the lion peeled off sharply, so close that he showered us with sand, and disappeared over the bank. I looked round to find Joan clutching a heavy saucepan

in each hand, white-faced but looking ready for anything. Her glasses had slipped down her nose and I pushed them back up for her, in the start of what would become an amusing ritual for us, when – with her hands full, holding a snake, covered in mud or flour, or inside a changing bag – she would thrust her face forward for me to push up her specs.

At dawn the Tiva river bed was silent, bereft of activity and littered with great quantities of elephant dung with its strong herby fragrance. As the first fingers of sunlight began to push through the bush, the sky would suddenly come alive with sound and movement as doves and sandgrouse flew down in their thousands to drink. The sandgrouse dropped steeply out of the bright sky with wild and haunting calls. After just two or three dips of their bills, they were off again in a sudden clatter of wings. The male birds also dipped their bellies, where a patch of specialised feathers soak up water like a sponge. After a fast flight of several miles back to the nest they would be greeted eagerly by their chicks, which drink from the males' still-soaked feathers. Within half an hour the flocks dwindled and we sat on the sand to have our breakfast, facing the bank so we would see any approaching animals.

One morning we were eating in silence, listening to the liquid calls of morning warblers, when a dark, cool shadow passed slowly over us, bringing a chill to both skin and mind. A big bull elephant had come up the river bed and walked behind us, just twenty feet away, a great moving wall of wrinkled skin, stained red with Tsavo dust, his heavy curved tusks skimming the sand. We didn't move and he didn't see us as he ambled forward with only water on his mind. Elephants move incredibly quietly on their big soft feet, especially on sand. Joan and I sat transfixed, wondering why we had not heard him coming. Then we both realised the reason and with minimal movement mimed our conclusions: Joan by wiggling her cereal bowl and me with an exaggerated chewing action. We had been eating cornflakes and the crunching in our heads had drowned out what little noise the bull had made. As soon as it was safe to speak, Joan lifted her bowl high and said, 'Tomorrow – porridge!'

The drought of 1960–1 is one of the worst Kenya has ever experienced. Hundreds of rhinos and smaller species died lingering deaths in the Tsavo National Park where we were working at the time. The park looked like a First World War battlefield: an alien planet of blood-red dust, shattered trees and enervated animals. The only movement was the tall, slowly spinning columns of red dust that marched across the ruined land, choking the small groups of animals that huddled listlessly in what little shade remained. Rivers, springs and waterholes dried up, and the concentrations of animals visiting the remaining water ate all available food within miles. In those dry conditions rhinos need to drink every day. Tied to the few remaining waterholes, the rhinos simply had no food within reach and would chew on dead branches, some of them dying from punctured stomachs.

Elephants are more mobile than rhinos, but still many died in that year, and several times we found an elephant – once a huge bull, his upper tusk pointing at the punishing sky – lying out in the burning sun, too exhausted to move even its trunk or, more important, its ears. Elephants have very large blood vessels running into their ears, which they flap to cool themselves. The bull we encountered would have lain there baking for another day or more before dying, helpless as hyenas chewed at his tail or the tip of his trunk. I carried no gun, but I knew that the skin at the back of those ears is as soft as fine suede, and with a sheath knife it was easy to sever the main artery so that the exhausted animal could die quickly and quietly. Leaning over that great noble head, just inches from a wise, unflinching chestnut eye, and releasing a torrent of several hundred litres of scalding blood, I felt like a murderer, but I also felt sure I was doing the right thing.

At times like this, when you come across an animal suffering horribly from a natural disaster, it is difficult to know what to do. I base my decision not on any moral compass but on what the next day or two is likely to bring to that creature. If it is a gazelle or zebra with a broken leg or disabling wound it will soon be noticed by a predator, brought down and die a fairly quick and natural death. But larger animals are sometimes found suffering horribly and have little chance of being swiftly dispatched. On the banks of the Nile I once found a hippo that had got stuck in drying mud with just its back and head showing. It must have been there for days and was horribly sunburned and blistered. The two or three hyenas that had found it could not get a purchase on its great rounded back, but had eaten away the wretched animal's protuberant eyes, its ears, nostrils and lips. Its ragged breath bubbled from between exposed teeth in the front of its mutilated head. I couldn't imagine the agony it had endured and was thankful that I had a revolver with me so I could end its suffering. On another occasion I came upon a buffalo that had dug a foot-deep circular ditch by scrabbling for days in an endless circle around one broken hind foot, which it had caught in the exposed roots of an acacia. He had finally been discovered – still alive – by hyenas. Another time we found a giraffe that had trapped its head in the twisted branches of a tall tree and could not pull it out. It had struggled there until being eaten alive by lions. Only the hind legs were eaten and from the amount of blood spilled it was obvious that the animal had been alive when found by the lions – predators will not risk injury and waste valuable energy killing a big animal if it is already helpless. I am keenly aware that nature does not care about suffering. There are wasps whose larvae eat caterpillars from the inside, ants that swarm over baby birds and eat them alive. I once watched in horror as a water scorpion with a needle nose injected a shrieking frog with a poison that slowly dissolved its insides. The bug then sucked out the soup through the same needle, leaving just a flaccid frog skin. Of course, these creatures are innocent – they know no other state – and the agony of individuals is of no concern to mother nature, yet it is hard to share her dispassionate view.

And how to react to injured or orphaned baby animals, cowering, bedraggled and frightened, in the long grass? I know full well that I am a total softie and the sight of a helpless and needy cub, kitten, foal, pup, chick or fawn brings out the mother hen in me. Into my shirt it goes and to hell with the sensible arguments that the animal has failed its Darwinian exams and should be left to die. There is a moral dilemma involved, of course. Especially if adopted at a very early age, many animals will imprint on their human benefactor and will grow up more attached to humans than their own kind. Then, even if raised free and in their natural habitat, they are ever after a creature between two worlds. But sometimes, relieved of the pressures and rivalries of growing up among their own kind, they develop interesting personalities that they would never normally exhibit. Still, the hope is always that, once fully grown, the animal will be able to return to the wild. If raised where they can gain experience of their natural habitat, they are often able to segue between their human 'parents' and their natural kin. Once grown, they can head off to pursue a full life in the wild, but maintain contact with the humans who raised them.

George and Joy Adamson achieved that end with all three of Africa's big cats and many other people have managed the same with different species. Joan and I raised several animals that successfully made the transition, including a wonderful serval cat who, though firmly imprinted when she first opened her eyes and saw me, went on to go wild in the Serengeti and twice brought her kittens to meet us. By picking up that dishevelled and hungry orphan, covered in ticks and sticky burrs, you are at least giving it a chance at life and I find that an inescapable imperative. As I write, a tiny orphaned warthog is nestled inside my shirt. There are little grunts of contentment – and that's just me.

For the BBC series *On Safari* Armand Denis wanted two half-hour films from me, one on the drought and another on the poaching that was going on in and around the eight-thousand-square-mile Tsavo Park. The Galana river, the major artery that flows through the park, had seen poaching on a massive scale that had been going on for years. WaKamba hunters – the tribe of my boyhood companions – concentrated on rhinos, which they killed with poisoned arrows, and leopards, which they caught in snares or gin traps that clamped their spiked metal jaws on to a foot. Large double hooks were put out at night in the river, baited for crocodiles, which they skinned right there on the bank, using the cadavers for bait the following night. Tsavo's Wardens, David Sheldrick, Bill Woodley and others, had mounted a campaign that took several years to bring the worst of the poaching to an end. Most of the elephants killed were being taken by hunters from a small tribe called the Wata (then known as the Waliungulu) who lived to the east of the park and for whom elephant hunting was the only way of life. Ian Parker, an old friend who became the authority on the tribe, believes their elephant-related culture goes back thousands of years. It developed because they lived among one of Africa's largest elephant populations that was close to the Indian Ocean and the trade routes to ancient civilisations with their hunger for ivory. Certainly the weapons with which they hunted indicated a lengthy period of evolution and refinement: the Wata bow requires extraordinary skill, and to an uninitiate can be the most humbling physical device. Almost six feet in length, and wrist-thick at its midpoint, it defies even the strongest of men to pull it more than a few inches, but a Wata of medium build points his arrow at the sky, pushes the bow forward as he lowers it and, spreading his shoulders, pulls the string back to his ear. The highly regarded English medieval longbow – the armour-piercing weapon of mass destruction used at the battle of Crecy in 1346 – had little more than half the pull weight of a big Wata bow. Each Wata hunter crafted his own bow and tailored it to his height, reach and strength.

Their arrows were works of art. Over three feet long and sturdily built, they were uniquely fletched with four vulture feather vanes – other tribes have three. The metal arrowhead was fitted to a six-inch shaft smeared with a deadly poison, usually supplied by the Giriama, a coastal tribe who boiled up the bark and chips of the acocanthera longiflora tree until it resembled tar. This poison-covered 'warhead' in turn fitted into a socket in the main shaft. When not in use the arrowhead and poison shaft were protectively wrapped in a strip of soft dik-dik skin and the arrows kept in a large leather quiver. The Watas' chosen target was the elephant's spleen and liver, well back on the left side of the belly, and the chosen distance about ten paces from the animal. A well-placed arrow could bring a bull elephant down in less than a kilometre and kill it in less than an hour.

The Wata were impressive men, and I was more interested in their history and their hunting technique than in their role as poachers. My aim was to film the stalk and the release of the arrow, so we produced some harmless props by fashioning arrowheads out of a soft solder material, which crumpled on impact and could be fired at an elephant causing no more than alarm. As long as the wind is right, there are no dry leaves underfoot and the animal is not part of a widely spaced group, stalking to within ten paces is not particularly difficult. But getting that close and then giving the elephant a sudden shock was a recipe for several frightening chases through the thorn bush with much laughter from the Wata, who could run much faster than I could and would always be well ahead. Considering the elephant's long and grim experience of men and arrows, we were lucky not to have provoked more serious reactions – many Wata hunters had been killed in similar situations.

It was exciting work that yielded some good footage and left me with the greatest respect for these men and their skills, and the wish that I had done much more to record their unique way of life. Their feelings for their prey, and for ivory, was evident when I watched some Wata men examining a large bull that had died during the drought, speaking in respectful whispers, measuring first the girth of a tusk with their hands, then carefully walking an arm, elbow to fingertips, down its length from root to rounded tip. Then they stood in silent, reverent admiration, a manifestation of an age-old bond between hunter and hunted that went back beyond knowing.

During the anti-poaching campaign in Tsavo a considerable percentage of the adult Wata men were captured and imprisoned, but in the process the wardens had come to admire them and sympathise with their situation. Realising that they were destroying this fine tribe, who knew no other way of life, Tsavo Warden David Sheldrick and Ian Parker proposed a scheme that would allow the Wata legally to hunt a small number of elephant, marketing the meat and valuable hides, while the ivory would go to the government. After all, with no valid justification apart from the income they brought to the country, visiting sportsmen were allowed to kill elephants on licence, so why not the people who had done so for millennia? This would allow them to continue their traditional way of life, while acting as guardians to keep the rest of the herds safe. Ian Parker made himself very unpopular among the hunting fraternity and the bunny-hugger wing of conservationists by energetically pushing through this project, but sadly the Wata found it almost impossible to adapt to a different way of life, particularly to using a rifle instead of their bows. Within a few years political interference and a short-sighted government led to the collapse of the scheme. For a while, the best Wata hunters found work with white professional hunters as trackers or gun bearers, but then, in 1977, hunting was universally banned in Kenya and this small, unique tribe has all but died out, along with most of the great elephant herds they once pursued.

4

My Mother and Other Animals

Our first two years of marriage were filled with a variety of projects that took us around Kenya, Tanganyika and Uganda. On one of those trips we stopped for coffee at the Bell Inn, in the little village of Naivasha. While waiting for our order I picked up an old newspaper and saw a rental ad for a house on an eighty-acre farm on the shores of Lake Naivasha. After my time filming lily-trotters there I had very fond memories of the tranquil lake with its abundant bird and fish life, its acacia-fringed shores with sleeping hippos and herds of wildlife, surrounded by distant mountain views of the Aberdares and the Mau escarpment. The farm was only five miles from the inn, so we gulped down our coffee and raced out to see it.

The house was a sprawling colonial-style building with a huge veranda looking out over the water. A pair of fish eagles were nesting in the giant acacias in front of the house and a group of lovebirds were burrowing into the side of the nest. On the lake shore, which had been trampled by hippos, we found a large green snake and many lily-trotters stalking over acres of purple water-lilies. It was love at first sight. We walked to the nearest house a couple of hundred yards away to ask if we could use their telephone. They wound the handle on the ancient party-line phone, and before long we had made the call and come to an agreement to rent it from the owner. It was ours! We couldn't afford it, but the house and setting were irresistible and as we drove away the fish eagles called out to us with wild yelping cries as if in congratulation and welcome.

Between safaris, to send off film for processing, repair tents or vehicles and restock with food, we needed to visit Nairobi, where we stayed in a mud-built cottage on my mother's land in Karen. Since we'd moved back to Kenya so long ago, my folks had become estranged and my father seldom came up to visit us from Southern Rhodesia, so Mum was always pleased to have our company.

It was time to upgrade our equipment. And in 1961 I splashed out and bought our first new car, a Land Rover. Not long afterwards our old Landy was stolen from outside the cottage. It had been quietly pushed away from the house before being started up, so I came up with the cunning plan of fixing an inconspicuous chain from the back of the new Land Rover to the metal burglar bars on our bedroom window. So devilishly cunning was this plan that I failed to remember it when I drove away early the next morning. There was a jerk and a loud bang, and when I looked back, there, in a cloud of dust, was Joan in her jammies looking out through a four-foot-square hole in the wall.

Theft was increasingly common around Nairobi, and often impossible to report because, adding insult to injury, the telephone wires had been stolen. The 1893 report on the survey for building Kenya's railway and the accompanying telegraph line had carried this prophetic note: 'It is recognised that stretching iron wire across wide expanses of unsettled country, where iron wire is the staple item of barter, is inviting constant interruption of the line.' A hundred years later the wires are copper and interruption of the line is a major industry.

My mother was delighted that I had found such a wonderful woman to share my life, and she and Joan got on really well. Mum still knew very little about animals but was always excited by new creatures that we brought home. She was game for anything and often had them staying in my old bedroom. Helping to look after my snake collection had given her a good start. When I was at school I had kept about two dozen, from ten different species, that lived together in a large enclosure made of four-foot-high aluminium sheet, furnished with small bushes, rocks and a pond. Young African boys would earn a few shillings by bringing in mice, frogs and lizards, which Mum dropped into the pen, shuddering as the snakes raced to grab their chosen prey. She was also an expert siafu extractor. Several times a year our garden would get raided by these ferocious army ants that overwhelm and eat any creature that doesn't manage to escape. Caged chickens and rabbits, tethered dogs, even babies are at risk from these numberless biting hordes. In the siafu season I would check my snakes early every morning and sometimes would find them gathered on top of the bushes or in the pond, with siafu hanging on to them with locked jaws. I would have to catch each one of the distressed creatures, some quite deadly, and hold them while Mum and my sister Jacquie painstakingly removed the ants with tweezers. It was a very long job, delaying our departure for school and work, and I wondered what the bank manager made of Mum's reason for being late. 'The dog ate my homework' sounds a lot more believable than 'I was pulling ants off the puff adder'.

Later, while I was filming with Armand Denis, a large chimpanzee arrived for me to look after. As I was away, Mum felt she couldn't just leave the animal in the shipping crate until I returned so she opened it up and obeyed the instruction on the box, which said, 'My name is Emily, please give me a drink and a broom.' The moment she came out of her crate, Emily got to work cleaning the house. Pausing only to empty the fruit bowl and finish the milk and sugar on the tea tray, she swept everywhere, looking under and behind furniture with far more diligence than Olatu, my mother's aged retainer. She then found a cloth, wet it under the tap and went to work on the tables and windows. If Emily did much more, Mum was going to get into trouble with the labour laws, but when my mother tried to stop or entice her back into her box this potentially dangerous animal baulked and became threatening. Wisely, Mum left her to it and after several hours of hard housework an exhausted Emily looked around with satisfaction at a job well done, took a couple of Mum's best towels to make her box comfortable and curled up to sleep. Mum's only comment was, 'She forgot to straighten the pictures.' Emily was

with us for several years, and she and Mum became great friends. Mum would often let her out of her cage and watch her making a beeline for the house, where she would look around in disapproval at the state it was in and get to work with her broom.

A half-grown leopard cub also turned up one day for me to look after for Armand Denis. Fortunately I was at home this time, as on opening the box I got badly scratched before the animal accepted me and settled down. It was almost weaned off the bottle, so we had to find a supply of meat – preferably something with some bone and roughage – and Mum mentioned Kinyanjui, her mole man. Kenyan gardens are plagued by what are commonly called moles, but which are in fact mole-rats, a hamster-sized rodent whose great yellow teeth do massive subterranean damage to lawns and vegetables. The area around Karen had an abundance of mole-rats equal only to its abundance of garden-proud ladies and Kinyanjui, an ancient fellow, made a good living trapping these creatures with a clever low-tech trap involving a flexible stick, a wire noose and an empty tin can that he buried in the mole-rats' tunnels. It was extremely efficient and at fifty cents apiece Kinyanjui did extremely well, as did our young leopard, who feasted on this ideal diet. Mum could never understand why Kinyanjui cut the tails off her moles, until she learned that the Karen ladies-who-lunch did not want to view a pile of ripening bodies when the time came to pay, they were happy just to see the tails. Old Kinyanjui was getting paid for his tails more than once, and we later heard that he'd started doing even better by skinning the rats and twisting strips to make realistic tails. He went undiscovered for years, then everyone thought it such a good scam that he got away with it.

What is it? Is it poisonous? Should I handle it? Mum was tested in this way fairly regularly. On one occasion her lack of knowledge – exceeded only by that of a Game Department officer – saved a tricky situation at the airport. A friend in Zambia had caught a pangolin and, deciding that I might like it, had it airfreighted to Kenya for me. A pangolin is an ant-eating creature the size of a large cat, covered in hard shiny scales. It has a long tail that balances the animal as it shuffles along, with its short front feet and powerful claws just brushing the ground – I described it in one film as 'bumbling around like a clockwork artichoke'. In order to stop the smuggling of illegally held wildlife, the Kenya Game Department had set up a station at the airport where they could check on all animal and bird shipments. Mum was informed that there was a box there for collection – and inspection. My friend had sent me the licence and export permit for the pangolin by post, so it was quite legal, but I was away and the permit was sitting in my post-office box. Unfortunately the shipping box had previously been used for pets, so stencilled in large letters on the side was CAT, and this was causing some confusion. The Game Department officer responsible for checks that day was puzzled, he had never seen a cat like this, but the box said cat, so a cat it must be, and that would involve quarantine and a hefty fee. Mum had no idea what the creature was, but looking into the box said, 'That's not a cat, it's something else.' What that something else was she knew not, but realising she was not alone in her ignorance she was prepared to wing it.

The warden peered into the box where the pangolin was tightly curled up like a pine cone and asked, 'What are those things all over its back?'

'Well, they look like scales to me,' ventured my mother.

'If they are scales it must be some kind of fish,' declared this guardian of biodiversity, pleased with his inspection.

Mum saw she was on a winning streak. 'Do fish have to be quarantined?'

'No, absolutely not.'

'Is there import duty on fish?'

'No, they go free.'

'You're absolutely right, it is a fish, so I had better get it home and into some water soon.' The warden agreed that this was urgent and wished her well as Mum walked out with the box, the proud owner of Kenya's first feline, ant-eating fish.

For several years, while I was away on filming trips, she also looked after our much-loved and almost legendary male baboon,

Bimbo. He came to us as an orphaned baby, who was happy to play and wrestle, and was everybody's friend. For a long time he lived free and would conduct mischievous raids on the nearby

Karen *duka*, our only shop. The Asian shopkeeper would give Bimbo jet-lagged apples and limp carrots, but he often helped himself to better stuff, and when Mum went there for supplies she would be asked kindly to settle 'Bimbo's account'. As he got bigger these raids became more like pillage and his account mushroomed to an extent that he had finally to be kept on a long running chain in the garden. Then, as he matured, he began to climb the ranks of his human society by dominating – read biting – first our friends' children, then wives or girlfriends, and eventually our male friends. The bites were all warning shots, nothing very serious, but enough to establish his position in the hierarchy until finally, when fully grown, he challenged me. This time it was for real – I was his group's Alpha male, the position he now wanted – and he went for me with his two-inch canines. He was standing on his hind legs drinking water that I was pouring from a canvas water bag when he made his move. He lunged at me and I fell back and in defence swung the heavy bag at him, but it hit his open mouth and one of his teeth ripped the bag open, showering him with water and leaving me armed with a floppy piece of canvas. Fortunately he came up against the limit of his chain and could only just reach me. His teeth ripped off my shoe and I narrowly escaped a serious mauling.

After that I spent hours trying to appease him, sitting just out of reach and subserviently grooming him with a short stick, trying to let him see I accepted him as the boss, but that we could still be friends. But our relationship had changed, and thereafter he became completely intractable. My mother, as the top-ranking female in our troop, had always been treated with love and respect, but now in his frustration he turned on her too. He was only following the male imperative of his social species to work his way to the top, but in getting there he had become highly dangerous. If released into the wild, he would have had a hard time surviving as a loner and most likely he would be killed by a predator, by a stronger, fitter male, or turn to raiding crops and get shot. Rather than keep him captive, alone, frustrated and angry, I felt the only answer was to have him put down. He was very much part of the family and it was one of the hardest decisions I've ever had to make.

Many years later another of our home-reared baboons had a meteoric career. Fording a river in the Serengeti one day I had spotted movement in a thick patch of bush. I investigated and found a female baboon with horrendous injuries, almost certainly from a leopard attack. One arm had almost been pulled off, the muscle stripped down to the bone. Her back had been laid open by slashing claws, the wounds were suppurating and the poor creature stank of decay. She just sat there huddled, with her arms folded, her eyes filled with pain. Beyond fear, she made no attempt to get away. She was slowly dying a horrible death. I put her out of her misery and as she fell sideways a tiny baby tumbled out of her arms. It was cold and appeared lifeless, and I could now see that the mother's nipples were torn and bloody from the infant's desperate attempts to suckle those empty dugs. I took him back to the car to show Joan and put him on the Land Rover's hood while I washed my hands. We sat for a while wondering how the female had managed to get away from her attacker and how long she had been suffering before we turned up. Our thoughts were interrupted by a loud clang from the front of the car. The heat of the hood had revived the baby and he had fallen off the edge, hitting his head on the way down. He sat there looking up at us with dark imploring eyes, his lips puckered in a silent appeal. Oh hell, the last thing we wanted was another male baboon but, despite our experience with Bimbo, we could not resist this skinny new creature and with great misgivings we wrapped him in a yellow duster and took him back to camp.

It turned out to be an amazing experience with a very happy ending. Primate infants have to cling tightly to their mothers from a very young age as their mum needs her hands free to climb. A human foster mother doesn't jump around in the trees, but the baby clings on just as tightly and screams with insecure fright if put down. Neither Joan nor I was looking forward to this, but we found that Lucky, as we named him, was unlike any primate we had known. The deprivation and lack of attention that he had suffered with his dying mother seemed to have totally affected his behaviour. He clung on to us, nothing would change that, but if we put him down, as long as it was in a warm, secure place, he accepted it and didn't make a sound. If he was hungry he wolfed down his milk, but didn't scream or bicker when the bottle ran out. His behaviour was everything anyone who has raised a baby primate – including a baby human! – could wish for, and he grew into a well-behaved animal who just seemed grateful to be alive.

We were visited in camp one day by Professor Irvin DeVore, an American who had studied baboons for years and was on his way to South Africa for more fieldwork. He was amazed at Lucky's undemanding nature, but more so by the little sounds that Joan and I made to him and his responses. 'How do you know all those vocalisations?' he asked. 'And what do they mean? I have never heard any of them.' This was Professor Baboon speaking, with years of field observation under his mortarboard, but we were not surprised. There are so many small sounds and almost imperceptible facial expressions that animals use to communicate that are often revealed only to those who have raised and lived closely with them. DeVore was so keen to learn more that he fell in love with Lucky and asked if we would part with him. Trying hard not to sound hugely relieved, we agreed to let him go, and Lucky soon flew off with DeVore to South Africa as a full-fare passenger, and went on to live a comfortable life in American academe, another baboon legend in the making.

The animal Mum was most proud to have been involved with was a baby bongo antelope, the only one in captivity in the world at the time, which lived in her house for a year. A forest officer friend had called to say he had a baby bongo that had been found next to its dead mother who had been caught in a snare. He didn't want the responsibility of trying to raise it – did I want to try?

The bongo is one of Africa's most beautiful and rarest animals. It is a very shy forest antelope, and in Kenya is found only on the slopes of the mountain ranges. I was very keen to try to raise it, so I collected it from the forest and, as I was still living at home with my mum at the time, set the animal up in my bedroom.

Bottle-feeding young wild animals can be a very traumatic exercise. The teat has to approximate that of the mother and even then is often rejected. The first few feeds tend to be a battle of wills as you hold the animal tight and struggle to get the teat into its mouth, while it jerks its head around and clamps its jaws shut. When it does finally take the teat it can sometimes chew off the end so that milk floods everywhere. In the struggle, if any milk goes down the windpipe and gets into the animal's lungs it can contract pneumonia and die very quickly. If you do manage to get the infant to suckle, you must also ensure the milk mixture is correct: too rich and the animal can get diarrhoea and rapidly lose condition, too dilute and it won't provide enough nourishment, and again the animal rapidly loses condition. For several days you walk a knife-edge, feeding every few hours, day and night, warming the milk, and warming and comforting the animal. It is a physically and emotionally draining experience. And then comes that deeply satisfying moment when your charge eagerly takes the teat, sucks rather than chews and dribbles, drinks its fill and lies down for a contented sleep.

With an animal as precious as the bongo it was even more worrying. It had spent a day and a night in the rain standing next to its dead mother and was in poor condition already. It was a bad start and I was so tied up keeping her alive I did not report it to the Game Department – as I should have – until it had turned the corner and was suckling well. By now the news had got out to the trappers, colourful characters in Kenya in the fifties who made a very good living capturing animals for zoos. Animals of all species were in great demand, most of them easily lassoed within fifty miles of Nairobi. The three biggest operators all complained to the Game Department that, as this was the only captive bongo in the world, it was irresponsible to leave it in the care of an inexperienced twenty-year-old. They more or less demanded that she should be handed over to one of them immediately, then proceeded to lobby hard as to which of them should have her. Fortunately, the Chief Game Warden was a fair man. He was impressed that I had got the baby bongo through the first and most difficult week. I think he may also have wanted to make a point to the trappers – that they didn't make the rules – for he told me to keep up the good work and gave me a licence for Karen, as we had started to call her. After a couple of months, when Karen was well established and had graduated to eating sweet potato leaves and Lucerne, I was able to go off on filming trips, leaving Mum in charge of this treasure.

One night while I was away a leopard from the nearby forest spent a long time trying to tear a hole in the thatch to get at Karen, until Mum scared it off with a compressed-air siren that she had bought during the Mau Mau time. Unfazed, the leopard came back a week later and killed a young reedbuck that she was also caring for at the time. She found the half-eaten body in the morning, dragged it close to the back door and called the Game Department to bring a live trap so the leopard could be relocated out of suburbia. But before they arrived it had returned to her back door and taken off its kill into the forest. Leopards were common in the area at the time, with many a dog snatched off a veranda or hauled out of its kennel, and one even being dragged away on its leash out of the owner's hands, in a traumatic end to that evening's walkies. Then there was the time that Olatu, my mum's cook, left the door open one evening and after two hours of desperate searching he and Mum had found the world's only captive bongo in the glare of headlights on the main road. Fortunately Mum was like a mother to Karen and the frightened animal followed her safely back home.

After living with us for eleven months, I accompanied Karen by air to New York where she would do two weeks' quarantine before moving to Cleveland Zoo. On about day three I was visited by Bill Conway, the iconic director of the Bronx Zoo in New York, which had exhibited the first ever captive bongo back in the thirties. He pointed out that Karen was being well looked after by professionals and that I really wasn't needed for the quarantine period, so said he would arrange something for me to do. I thought this might mean a trip round his zoo, but the next day he gave me train tickets that would take me all the way to Jackson Hole, Wyoming. There I was met and taken to Moose, a tiny village nestled below the Grand Teton Mountains, the ruggedly beautiful range I recognised as the magnificent setting for *Shane*, that greatest of cowboy films. Here I was to stay in an old log cabin with Olaus and Mardy Murie. Olaus had just retired from being the first field biologist for the US Fish and Wildlife Division, and he and his wife could not have been kinder or more inspirational hosts. We spent days moving around the nearby Yellowstone Park, the Elk Refuge and many secret places the Muries knew, and I eagerly soaked up their encyclopaedic knowledge of the American wilderness and their gentle way with nature. They had spent most of their lives in the far north and had recently finished a survey of a vast wild area in Alaska that is now the Arctic National Wildlife Range, and they spoke warmly of a young student who had been with them, George Schaller, who would go on to become the world's finest field biologist – and a good friend of mine. When the bongo's quarantine was over, Cleveland laid on an incredible reception for 'The World's Only Captive Bongo', a brand-new building, opened by the mayor, with TV, press, parades, parties – you name it – but all that razzmatazz was totally eclipsed by those ten days in the company of two wonderful spirits of the Far North.

Many years later there were still only a few bongos in captivity. I was asked – almost begged – by the Milwaukee Zoo if I would try to catch more so they could start a captive breeding programme. My first thought was to contact Richard Gicheru, who had worked at Nick's forest station, and with whom I had tracked bongo several times. I would need the help of his forest skills if I was to succeed, but I had not heard of him for about twelve years, so it seemed a very unlikely proposition. However, when I got back to Nairobi after a long trip away a pile of post awaited me, including a letter from Richard – my old forest friend!

Bongo were being heavily poached in the sixties and capturing some to set up a captive breeding programme seemed to be a sensible idea. Arabian oryx and Père David's deer had been successfully brought back from the brink of extinction in similar operations, and though at the time the bongos' situation did not appear to be urgent, it now looks like the exercise may yet prove to be the species' salvation.

The Game Department wanted us to start in the area around the South Mathioya river on the eastern side of the Aberdare mountains. Several well-qualified people had tried to catch bongo over the years, but their methods had involved stretchy nylon snares, covered pits or lassoing the animal when it stood at bay to a pack of dogs. In each case they then shot the bongo with a tranquillising dart. But what do you then do with a stressed and sleeping three-hundred-pound-animal high in a cold, wet forest? They just hadn't thought it through. After being carried on a stretcher for hours, or tied up in the back of a helicopter, every bongo had died and the operations were halted. When Richard Gicheru and I recced the South Mathioya we agreed that we were dealing with very sensitive animals who routinely kept well away from human sound or smell, and that every stage of this operation must go *polé-polé* (KiSwahili for slowly). We planned to catch at about eight thousand feet, in the bamboo zone, where there were plenty of building materials for the traps I had designed. These would be high-walled pens, twenty feet long by five wide, with a big sliding door at each end. Richard and I scoured the area until we located a couple of trails that bongo had been using and there, with Richard's team, we built several traps. We then lifted up the doors and fixed them so both ends were open and left them for a couple of months until the traps were completely overgrown with creepers – no more than a dark tunnel through the bamboo forest.

We made camp about a mile away and set the traps, holding up the heavy doors with pegs attached to a stout bamboo trigger placed across the trap halfway along and high enough for smaller animals like bushbuck and duikers to pass underneath. When knocked down by an animal this trigger would dislodge the pegs and release the doors. Joan had been with us throughout, and every day she and I checked the traps early morning and evening, staying as far away from them as we could to avoid leaving our scent. Joan gathered and pressed a collection of bongo food plants and we sat round the fire in the evenings, sharing our maize porridge and stew with Richard and our crew, and quizzing them on the Kikuyu names for balsam, mimulopsis and other flowering plants. Every day I fished for trout and occasionally caught one, which we ate garnished with bamboo shoots. We measured our days by the *gathano* mists ghosting through the morning bamboo, the sudden icy showers that caught us walking to the traps, and the scent of cedar smoke in the clean, cold air as we crawled into our sleeping bags. Our nights were serenaded by the screams of tree hyrax, the liquid calls of Abyssinian nightjars, the distant crack of a broken branch or rumble of an elephant. On the morning of the twentieth day we saw the doors were down on one of the traps. We raced towards it, then crept up and put our eyes to the cracks in the bamboo walls. Inside was a beautiful young female bongo, unharmed and unperturbed in the shadowed confines of the trap. I'd had plenty of time to think about the next stage of the operation, and knew it would be a mistake to move too fast and perhaps lose the animal to shock. *Polé-polé* was still our watchword. Rain was on the way, but I had a well-rehearsed plan.

Working quickly and quietly, using bamboo that we had cut earlier and stockpiled nearby to avoid noisy chopping, we built a shelter, eight feet square, on one end of the trap. We carpeted the floor with leaves, made a pile of food plants in one corner and covered the whole hut with black polythene so it was warm, dry and dark. As soon as we lifted the door and the bongo saw that shadowy refuge, she walked in and lay down. An hour later we heard her eating, and there were huge smiles all round. Over the next couple of days we moved our camp closer to the trap, and began the process of getting her completely used to the sounds and smells of people. Every day we collected fresh food plants for her and before long she was feeding from Joan's hand. Next I brought up a sturdy plywood crate that we positioned in front of her door and put all her food in there. Soon she was going in and out of it happily. Then we closed her in for a couple of hours a day, and when she was relaxed with that we would lift one end of the crate and rock it back and forth to get her used to movement. By now we could stroke her, and our hands would come away stained orange and oiled with the musky-smelling lanolin that waterproofed her glowing coat. This all took about a month, but by then she was totally habituated and ready for her ride down the mountain.

We were four miles away from the forest edge where there was a steep but motorable track. Four miles, with a descent of about three thousand feet along steep muddy trails and across two rivers. Her crate had rope loops along its sides, through which we passed long, strong bamboos. I had taken on twenty-four men from a nearby village and with two teams of twelve we carried her on a five-hour trip out of the forest. Soon she was subjected to a new and unexpected noise, the Land Rover, but so far she had done the trip sitting down and she didn't bother to get up when I started the car. We called her Mwathe, the Kikuyu word meaning from the depths of the forest.

Mwathe was the first of over thirty bongo that Richard Gicheru and I captured over the course of several years. Some of them were magnificent males, dark, rich mahogany with heavy horns, who took longer to adjust and a lot more men to carry out. Six were pregnant females who all successfully bore their calves in the dedicated quarantine station that we built at our Lake Naivasha home.

The bongos went in breeding groups to some of the best zoos in the world, where they have flourished. Back then there were thousands of bongos in Kenya, which were not only being hunted on licence, but were also being heavily poached. Today there are over five hundred bongos in captivity – more than now exist in the wild in Kenya – and from this well-controlled world breeding programme some are being brought back to Kenya to try to re-establish them in the protected areas where they have almost disappeared. This is still a work in progress, but with luck will one day provide a satisfying finale to a bold venture.

Joan and I still stayed in Mum's cottage when in Nairobi, but now she enjoyed coming down to bongo-sit while we were away filming, and to renew her friendship with these wonderful animals that had started with her beloved Karen. She also enjoyed the birds and the fishing out on the lake, but one day, casting the large, multi-hooked lure that we used for black bass, she managed to hook it through her septum. With this four inch, brightly coloured plastic fish hanging from her nose she looked like some New Guinea native and I'm afraid we couldn't help giggling. She took it with her usual good humour and insisted I should bring out a mirror, which got her giggling too, before our neighbour Dr Bunny gave her a local anaesthetic shot and removed it – an operation he carried out on many weekends.

When she came down to stay at Naivasha Mum also looked after Sally, our tame hippo. Sally had been found abandoned as a pig-sized baby after a severe drought at Lake Baringo and she lived with us for eight years, growing into a five-thousand-pound big mama. She was free to come and go into the lake, but as our house had originally been built for an invalid, with extra-wide, wheelchair-friendly doors, Sally also had the run of the house. The size of a sofa, she would lie at Mum's feet by the fire, and we would often hear shrieks of horror from guests when they found Sally lying with her huge head on their beds, sucking the corner of a pillow. When excited, she would open her mouth wide and rock her head from side to side, exposing her foot-long teeth. When the lawn sprinkler was on, she would stand with her open mouth over it, enjoying a monster mouthwash. Sally was totally trustworthy, and visiting children loved shoving handfuls of grass and uneaten picnic scraps into her cavernous maw. She was besotted with our tame aardvark, a long-nosed, huge-eared, forty-pound anteater we named Million – from the line in the song, 'Aardvark a million miles for one of your smiles.' (Well, it seemed funny at the time.) Million would spend a couple of hours every evening digging huge holes in the lawn, supervised by our cook's young son, Babu, who was mentally handicapped and delighted to have a responsible job. While Million excavated, Sally would lie outside the hole, usually with her mouth wide open collecting the flung earth, waiting for her friend to emerge. Babu would watch patiently, leaning on his spade, then refill the hole and repair the damage, but our lawn still looked like a bombing range.

This wonderfully eccentric trio were a great feature of our garden for many years. Sharing the lawn with them was a male crowned crane that had come to us as an orphaned chick. He grew into a beautiful, slender bird, about a metre high, pale dove grey with a crown of three-inch golden bristles. We'd had no luck finding a mate for him and he was clearly frustrated. However, in the middle of the lawn was a metre-high standpipe, in leaden grey, topped with a shiny brass tap. A Giacometti crane! Every day our crane would dance to his slender, unresponsive love, leaping and pirouetting with spread wings and bobbing head. He would offer her gifts of grasshoppers, pushing them into the opening of the tap and picking them up over and again when she dropped them. Happily we eventually found him a proper mate and they went on to raise chicks in the reeds along the lake shore. He and the standpipe are still good friends.

Joan and I became well known for taking in any wild orphans, but not all the creatures that came to us were large. On one occasion, just after we were married, Mum got a call from the Nairobi Central Post Office to say there was a parcel too big to go in our box, and could she please come and collect it immediately as a large animal was breaking out of said parcel. (Posted items are not delivered in Kenya, distances are too great and physical addresses too obscure; we collect our mail from a numbered box at the post office.) Mum hurried the few blocks from her work at the bank to the post office and was led into the sorting area where a large and excited crowd had gathered around a shoebox, well wrapped and tied with string. The box was jerking around on the table and something was pushing up at a ragged hole in the lid. Every now and then the scrabbling would stop and the box would vibrate violently, sounding rather like a captive motorbike. The workers drew back to allow Mum to approach and stood in silent anticipation, waiting for her next move. Feeling like a gladiator being urged out into the arena, Mum edged a little closer. She could see from the label that the parcel had been sent by the Kakamega Forest Officer, who gloried in the name of Dick Pricket, which made her wonder whether she really wanted to see what was vibrating inside the box. Apart from providing a little light relief, the label did nothing to help identify the contents, but suddenly all was revealed.

Out through the hole it had chewed in the lid squeezed a huge black, white and brown beetle the size of an orange. The audience fell over themselves backing away, leaving Mum to *mano-a-mano* combat with a goliath beetle, Africa's biggest. The beetle swung its head from side to side, then doubled its already intimidating size by opening its heavy wing cases and vibrating its wings to produce the motorbike sound. The staff were now scrambling over piles of parcels to get away from this creature, which was growing more frightening by the second, as Mum was faced with the usual questions. What is it? Is it poisonous? Should I handle it? She knew from picking beetles off her roses that this monster was warming up for take-off, and that if she let that happen – and this helicopter gunship started to fly around the post office – it would probably require the fire brigade to bring it down. Resolutely moving in – doubtless while muttering rude things about her only son – she took a deep breath and simply grabbed the beetle and stuffed it back through the hole. It was as easy as that, but to the post office staff she was an instant heroine. As loud applause and cheering broke out, she was almost carried out shoulder high. When I got back from safari two days later the beetle was happily eating a banana in the bottom drawer of her desk in the bank. There is a heart-warming ending to the story as well. Some considerable time later, my mother suddenly collapsed in the same post office, vomiting blood from a perforated ulcer.

Some of the staff recognised her, carried her out to their car and delivered her to the hospital. When I went to thank them for what they had done one of them said, 'That lady was too brave, of course we had to help her – she saved us from that giant beetle.'

All these years my dad had been living down in Zimbabwe, pretty well distanced from us, but now he had retired from Liebig's to become a sought-after and well-paid consultant for the World Bank, and travelled all over Africa evaluating livestock projects. He visited great, gleaming factories of tiles and glass that he would have given his eye teeth to run when he was younger, but actually they had been hurriedly built at huge cost with international aid money, sometimes hundreds of miles from railhead, electricity, or even a sustainable supply of cattle. After assessing several such white-elephant projects, where his reports could only point out why it wasn't going to work, he gave up.

Before long he and Mum were back together again, which was great. They did the usual stuff, cruises to the Orient and so on, but Mum was happiest in her little house where I still brought her the odd animal to look after – the odder the animal, the more she liked it. The last time I saw her, there was a giant fruit bat hanging above her sideboard, which crawled down into the fruit bowl when it felt hungry, and six crocodile eggs nestled in a sandbox in her airing cupboard. A week after that visit I flew back to Nairobi from the Serengeti and was met by David Mutaba, the engineer who looked after my plane. 'I'm sorry to be the one to tell you, Alan, but your mum died yesterday.' An asthmatic seizure had led to a heart attack. Mercifully, my father had been with her. She was buried in the cemetery just across the road from the Nairobi National Park animal orphanage and partway through my eulogy I heard Sebastian, their large male chimpanzee, start up one of his long, loud calls. I stopped speaking and I know Mum would have loved the scene: her gathered friends smilingly listening to the wild, hooting chimpanzee farewell. Like the 'Last Post' played for a soldier, it was a moving and fitting valediction.

5

Balloons and Other Signs of
Madness

In 1961 Joan and I met the English author and adventurer Anthony Smith who, with little effort, had persuaded me to be cameraman on something called the *Daily Telegraph Balloon Safari*. Tony was the *Telegraph*'s science correspondent. Tall, balding, erudite, brimming with eclectic learning, enthusiasm and wit, he would have been a marvellous companion on any expedition, but on this one he was crucial to its success. He had conceived the wild notion that it would be good to learn to fly a hydrogen balloon, bring it to Africa and balloon over wildlife. (Tony is still a man of wild notions. As I write, he is eighty-four years old and has just crossed the Atlantic with three other old age pensioners, on a raft made of large plastic pipes called *An-Tiki*.) The timing was important, as 1962 would be the hundredth anniversary of Jules Verne's first scientific adventure book, *Five Weeks in a Balloon*, a story about a ballooning expedition across Africa, whose fictitious leader happened to be the science correspondent for the *Daily Telegraph*.

In 1961 there was neither balloon nor a balloon pilot in England, so Tony had bought a balloon and done his flight training in Holland. Before he completed his training he landed very heavily in a frozen ploughed field and his two instructors – the only ones in Europe – were thrown out and landed badly, putting both out of action and temporarily unable to make

110

another flight. The Dutch Aero Club decided to grant him a licence provided he passed his theory exams – possibly because they knew he wasn't going to be flying in Holland – rather as the Russians back then would give you a degree in surgery provided you didn't wield your scalpel in Russia. The *Daily Telegraph* editor liked the symmetry of retracing Verne's hero's journeys and agreed to sponsor the project and run a series of articles on the expedition. A publisher had said yes to a book and the BBC wanted five half-hour films.

The first flight was to be from Zanzibar to the mainland and on 30 December 1961 I flew there to join Tony and Douglas Botting, who would be our stills photographer and extra muscle. Tony had pulled it all together just in time for the centenary. The timing was perfect, yet in another respect it was dreadful, for 1961–2 was what we now know as an El Niño year and was the wettest that century. East Africa's roads, bad at the best of times, were in the process of entropy – washed away, closed by landslides or damaged bridges or just so deep underwater as to be impassable. For every flight we needed two five-ton trucks loaded with hydrogen cylinders. It was clearly going to be a lot of fun. There were cyclones off Mauritius and the contrary winds that accompanied the storms made a mockery of all the predictions upon which we'd based the various flights. I think it was about now that I started quoting the saying that if you want to make God laugh, tell him your plans . . .

Tony had called the balloon *Jambo* (KiSwahili for hello). The flight from Zanzibar to mainland Africa went smoothly, and had the added distinction of carrying the first, and I'm sure only ever hitch-hiker on that route, a young South African student, Charl Pauw, who went on to great things on television down there. Heady with that success, we headed for Lake Manyara National Park in northern Tanganyika, a four-hundred-mile drive on flooded roads that took us three days. The park is a long, narrow strip of forest and woodland between the thousand-foot-high Rift Valley wall and the lake, and we camped just outside the park beneath the escarpment. We had been experiencing a big storm every day by now, but always in the late afternoon, so for our flight into the park we started the inflation as early as we could, but with just a few unskilled locals to help us it took longer than expected and we were not ready until late morning. We could see thunder clouds building, but we expected to finish our flight before they broke. As Joan left to drive to the top of the escarpment to watch us, we climbed into the basket still feeling fairly confident.

The flight started out perfectly and soon I was busy filming while Douglas took stills. We stabilised at a good height for filming and flew along the lake shore parallel to the escarpment. To our delight the buffalos, zebras, elephants and giraffes that passed below us took no notice of this small, slow, silent cloud. Most animals see in some version of black and white anyway, so our orange-and-silver colour scheme would not have worried them, but I felt it was our size, shape, slow movement and especially the total silence – unlike any bird of prey or aircraft – that failed to trigger a fear response. It was an extraordinary experience, but not one devoid of worry. The basket and its supporting ropes – slender hemp ropes, not cables or nylon – seemed minimal for the task of keeping us safe a thousand feet above the earth and the balloon itself, with its little wooden flap of a release valve on the crown, seemed totally inadequate. We were all aware of the fragility of this voyage and I found it best to look down in wonder rather than look up and wonder, what if?

Given the weather we had been having it was not surprising that the small cloud we had dismissed on take-off had grown. Grown? Hell, it had mushroomed into a cunim, a full-blown cumulonimbus, which now towered to maybe thirty thousand feet ahead of us. This was the kind of cloud for which the instruction manual tells you firmly to 'land and deflate at once'. But we could not land as we were now over the lake, and soon we moved into the cold umbra of that cloud and began to rise in the air that was being sucked up into its dark heart. In a few minutes we were a mile above the earth, so high that the thousand-foot escarpment became just a small crease in the map of Africa. I noticed that Tony's hand, which was holding the altimeter, was shaking uncontrollably. He kept changing hands but the trembling was transferred along with the instrument. I realised that this was only his second flight outside the flat fields of Holland, and that he had got his licence by putting his examiners in hospital. I knew he had never been anywhere near this high nor ever been this severely tested. I tried to think of something else and busied myself finding a new carton of film, and as I opened it the thick cardboard container was whisked out of my hand and went spinning up into the gloom and out of sight. We were being sucked upwards by the same forces and only our weight

was slowing ascent. We could have released more gas and so reduced our lift, and stopped climbing. But then we would be dangerously heavy, held up only by the cloud's powerful column of rising air. And eventually we would move away from the column, or it would dissipate, and then we would drop so fast that throwing out our supply of sand would not be enough to arrest our fall. I did not understand all the mathematics of balloon load and lift, but I had flown enough to know that if we went up into the underbelly of the black and growling mass above us we would be torn apart and spat out in many pieces. And I knew that my pilot, far more aware of balloon dynamics and behaviour, was probably even more scared than I was.

At nearly eight thousand feet we stopped climbing. We sat there in the cold twilight for several minutes just below a ceiling that looked like an upside-down North Sea in a gale, a tumult of black waves that tossed and rumbled as loudly as any ocean. We were totally silent. Then, almost imperceptibly, we started to move. If we moved upwards we would soon be just another cautionary balloon statistic. But no, we were going sideways! We were moving out over the lake and slowly we emerged from under that cloud-beast and out into the sunshine. Seen from the side, the thunderhead had morphed from a threatening black hell to a photogenic pile of sunlit ice cream. Suddenly we were talking again, our relief palpable, and in textbook fashion we crossed thirty miles of soda lake and landed softly and safely near the potholed ribbon known, laughingly, as the Great North Road. We were soon picked up by Joan who had been watching, as terrified as we were, through binoculars from the top of the escarpment. From her position the balloon had disappeared from sight under the storm cloud. It must have been desperately worrying for her.

Balloons are admired for their slow, stately progress through the skies, but they also bring that same unhurried measure to their peril. I have been close to death a number of times, several of those in flying machines, but only a balloon gives you the time to consider at your leisure what form your death might take and how long it might be before you find out. Chewed up and spat out by that great storm cloud? Surviving the turbulence but dying from cold and anoxia at thirty thousand feet? Having your spine shortened by six inches on impact after a long, long fall? How about something lingering, like landing in a concentrated soda lake five miles from shore, or maybe something instantaneous like lightning exploding the huge bag of hydrogen just above your head? Only a balloon can put you in such clear and present mortal danger and then allow you to take your time, make your plans: God is laughing.

On another flight the local newspapers prematurely reported that we were dead when our balloon was seen to go down vertically into the Ngorongoro forest. After taking off from the Nairobi Air Show in high winds we experienced another eleven-thousand-feet-high dance with death, but managed to end the safari with a perfect, low-level flight over the Serengeti herds. Our films were a hit, though sadly the BBC in its wisdom had given us only black-and-white film, which for an orange-and-silver balloon flying over Africa – as colour TV dawned – seemed somewhat less than visionary.

Anthony Smith's hydrogen balloon had been an extraordinary way to see the world and I loved it, but the cost per flight was astronomical. Ten years later, on a trip to England, Tony introduced me to the new hot-air balloon that was so much cheaper and safer than *Jambo*, the bag full of expensive and explosive hydrogen that we had flown in together. The hot-air balloon operated on the simple fact that hot air rises and was powered by a couple of gas cylinders. Tony felt it would be a good filming platform and had arranged for me to have a flight. As we lifted off I leaned out of the tiny basket, framed a single daisy with my fingers and imagined a film shot. We rose slowly and my frame took in more daisies, the village green, the village itself, then the rolling Oxfordshire countryside in what would have been one long, smooth take – unobtainable in any other way.

'I want one,' I called down to Tony.

'I knew you would,' he shouted back. 'I've already ordered one for you!

When it arrived in Kenya we used the balloon to get low and slow shots in several films, and it was a source of great enjoyment making early-morning flights with friends. In the late sixties, Joan and I had a small safari company called Africa Thru the Lens, and every year we made a couple of month-long trips with a group of twelve Americans. They were usually interesting people but we were happier in the bush on our own, and as I now had plenty of film work we were looking for a way out of having to make these trips. My friend Richard Leakey was in a similar position. He had run a successful camping safari operation for some years to help finance his fossil-hunting activities, but now that he had proved himself by making several spectacular finds, his work was being funded, so he too wanted to pull back from his safaris. We decided to combine our operations and employ a friend, Jock Anderson, to run the company and do the guiding. Imaginatively, we called ourselves Root & Leakey Safaris (known in the business, of course, as Rotten Lucky Safaris) and for some years, until we tired of the crowds and the growth of low-end tourism, we led the field.

One of our clients was President John Kennedy's widow, Jackie, who, when she heard about our balloon flights, arrived with her children John and Caroline to stay with us at Naivasha. The balloon would only carry three, so at dawn I took off with Jackie and Jock Anderson to fly over the lake. Flocks of snow-white egrets exploded from the feathery papyrus islands, and thousands of coots skittered ahead of us, unzipping the glassy surface with their long take-off runs. A fish eagle circled the balloon as it floated through his territory, and I swooped low over the vast beds of water-lilies so Jock could lean over the basket and pick Jackie a purple blossom. It was a glorious morning, as only the

116

Kenyan high country can deliver. As I approached the landing area I could see that we would pass through the smoke rising from a large grass fire. I was not concerned, and let the balloon cool and start its descent. I had allowed it to come down faster than normal as I anticipated getting some lift as I passed through the hot smoke. Unfortunately, I still had little experience of flying the balloon, but had racked up a couple of thousand hours flying a standard fixed-wing aircraft.

However, there is a crucial difference: a plane will get lift as its broad wings pass through rising hot air, but a hot-air balloon is kept aloft by being hotter than the surrounding air, so as I flew into the smoke and the temperature differential fell, so did we – like a stone. Telling Jock and Jackie to get down in the basket, I gave a long, desperate burst on the burner, but it was too late. Just clearing the burning grass, the basket swung across the Nairobi–Nakuru railway line and plunged into the middle of multiple telephone wires, which run extra high in Kenya to allow giraffe to pass underneath, where the basket got trapped lying on its side, the flaccid envelope rapidly collapsing. Soon the wires started to snap one by one with a loud twang like breaking guitar strings. A few more twangs and we dropped the remaining ten or so feet into a fine crop of maize. Jackie landed heavily on top of Jock and I completed the sandwich, and as we crawled out of the basket the Nakuru train thundered by just twenty feet away. Jock had cuts and bruises, and a story he tells to this day. Jackie had a badly sprained ankle that swiftly swelled and coloured, but she took it all with the easy grace that was her trademark. Kenya's President at the time, Jomo Kenyatta, spent his weekends at Nakuru, and the joke soon went around that when he found his phone didn't work that morning he was told, 'Sorry, Mister President, but Alan Root is on the line.'

Despite my obvious lack of experience I now decided I would like to balloon over Mount Kilimanjaro which, at 19,340 feet, is Africa's roof and the highest solitary mountain in the world.

Every day I contacted the Met Department and used the wind patterns to pinpoint the best launch site for passing over the summit. And so it was that after weeks of planning and test flights, we set out to camp in a forest clearing near Tarakia, on Kilimanjaro's eastern flank. We were a skeleton crew, just Joan and me and Giles Camplin, who had come out from England to help. In order to generate the added pressure we needed from our six butane gas cylinders we stood them in a circle round a small gas cooker, then covered them all with an asbestos blanket. (Propane has a much higher pressure, but a tank left out in the midday sun could explode, so it is unavailable in equatorial countries.)

On the day of the flight we were up at five and as the sun came up I was suddenly aware of the reality of what I was about to do. Kilimanjaro loomed massively above us, a ragged grey cloak of cloud pulled loosely round its huge shoulders and closing in: for the first time the mountain filled me with fear. I knew that ventures such as this build their own momentum and become almost impossible to cancel even when conditions are obviously not right. Pressured by the media and crowds of well-wishers, a good friend of mine, Malcolm Brighton, and a young American couple who were sponsoring him, had recently taken off in less than ideal conditions to try to balloon across the Atlantic. They went down somewhere in those freezing waters and disappeared without a trace. I was glad to have none of those pressures. Old friends Ian Parker and Alistair Grahame were in the air at first light in Ian's plane, with a camera fixed on the wing. They had radioed us that up above the cloud the weather was perfect, so was the report on the winds aloft – we had a 'Go'. I had planned all along for a solo flight, but as I tested the temperature inside the balloon I realised that I could carry a little more weight. I knew that Joan desperately wanted to come along, so given that she was already warmly dressed I asked casually, 'You ready?'

Joan often said that it was more fun, and less stressful, to join me on dangerous missions rather than watch and feel powerless – she was in the basket in seconds. On the north-east trade winds we climbed steadily up the mountainside through dark, damp cloud, the occasional ghostly skeleton of a dead cedar drifting below us, until after twenty minutes of flying blind we burst out into glorious sunshine. We were at the foot of the sheer four-thousand-foot gorge of the Great Barranco of Mawenzi, the eroded lava plug that is the lower of Kilimanjaro's two summits, a crumbling hulk of rust-coloured, seldom-climbed cliffs flaming in a morning light that threw purple shadows into the ravines. To the west, jutting out of a flat layer of clouds, the rounded dome of Kibo, the summit, crouched under the weight of its glaciers. We were on our way.

The Met Department had told us the winds would veer steadily as we climbed, until, at twenty-two thousand feet, they would blow east-north-east at fifteen knots. Sure enough, when floating a mile above Mawenzi's jagged peak, we swung notice-ably to the right and began a perfect traverse of Kibo's icy roof. To get a wide view of this vast panorama I had climbed to twenty-four thousand feet. Ian made several passes in the plane, nose-high, the engine labouring in the thin air, and I was filming from the basket, so we were getting plenty of good coverage. Joan had been taking stills, but I noticed she was behaving strangely and, with shaking hands, she was having trouble reloading her camera. Several times, looking down, I had been hit by a surge of fear at the precariousness of our position and I imagined that despite her usual courage she was simply scared. I reassured her that everything was OK – she needn't be fright-ened. 'I am *not* frightened!' she shot back indignantly, 'but there's definitely something wrong with me. I think my oxygen isn't working, I feel awful.' Sure enough, her oxygen hose had become disconnected and she was showing the lack of judge-ment and coordination that are the signs of anoxia. I pushed her glasses back up on her nose, hooked up her oxygen again

119

and within a couple of minutes she was fine and gave me a well-deserved and scornful, 'What do you mean, don't be frightened!' I couldn't have asked for a better basket-mate.

About two hours after lift-off, and four miles above sea level, we gazed down from our gossamer craft as the dome of Kibo, the highest point in Africa, slid slowly by – almost a mile below us. Kilimanjaro is a dormant volcano and the perfect crater gracing its summit is still warm with the memory of its last eruption, perhaps just a few centuries ago. Any snow falling on it soon melts, but the mountain's broad shoulders are covered with glaciers and snowfields. I lost height to sweep across the bleak Shira plateau and down across the forested western slopes, but then the balloon came slowly to a halt. I don't know what combination of terrain and atmosphere created that dead spot, but we were totally becalmed over a sheer-sided gorge, and while for a short time it was an interesting experience, we were burning precious fuel and getting nowhere. We still had fifteen miles of forest to cover before we could make a safe landing – we had to climb back up to the more reliable winds at sixteen thousand feet or more.

By now heavy cumulus had filled the skies above us, but we climbed through them and as we emerged into sunlight the balloon heeled over sharply in a good wind. This second climb had used a lot of fuel and I was down to my last cylinder, which by now was cold and had lost a lot of pressure. Instead of strong and blue, the flame was orange and fluttered weakly. When these aluminium flight cylinders are empty, the fuel gauge float rocks back and forth with an eerie squeaking sound. Every slight movement in the basket brought forth a worrying lament from five of our six tanks, reminding me of how far we had to go, and how little fuel we had to take us there. Fortunately the wind had picked up, and after less than an hour we saw, through a break in the cloud, that we were almost over the forest edge and could begin our descent.

Given my fuel was so low, I decided to try for the 'Montgolfier effect', which describes the beneficial warming allegedly provided by air friction when a balloon is allowed to fall at high speed. In 1783, at Versailles, the Montgolfier brothers had sent up the first passengers ever to ride a balloon – a duck, a cock and a sheep. Upon landing the cock was discovered to have a broken wing, and the brothers immediately declared that leaving the bounds of the earth could fracture bones. Why broken-winged birds did not regularly fall from the sky was not considered, nor the fact that all three passengers had landed while crammed together into a small cage, which might have explained the injury. It was hard to put much faith in the 'effect' named after the pioneering Montgolfiers, but with so few choices and so little fuel I had to give it a try.

We were two miles above the coffee and maize fields of Sanya Juu, which was a long way to fall. Our descent speed built up and at fifteen hundred feet a minute the air flowing over the segmented shape of the balloon may have marginally warmed us, but it also caused the balloon to rotate rapidly. We hung on to the wildly gyrating basket and had our first really frightening moments of the flight. Over open farmland and still slowly spinning, I released the trail rope. This is a very heavy hundred-and-fifty-foot-long rope that hangs from the basket and acts as a brake when coming in to land. It fell badly and tied itself in a large knot, so I laboriously hauled it all back into the basket to untie it and try again. I started to release it a second time, hand over hand, but I was physically and mentally exhausted, and decided to let it just fall. As the heavy rope whipped up from the basket floor with increasing speed, I leaned back away from the flailing coils, not noticing that the rope passed behind my back and up to the burner frame. Suddenly it snapped tight like a gigantic whip across my back, sliding up to hit the back of my head, flinging me forward and half out of the basket. I hung over the edge, watching my glasses spin towards the ground. Joan had grabbed my shirt and I turned to see her horrified face switch to a grin as she offered me her specs. Like me, she was very short-sighted, but our prescriptions were not the same, so as we came in to land I was unable to judge distances accurately and I asked her how high she thought we were. 'Don't ask me,' she replied. 'You've got the glasses – I can't see a thing!' I burned the last of our gas to slow our descent, but the landing was heavy. The basket ended up on its side, but we were intact, and were quickly surrounded by a crowd of curious and laughing Africans. I was kneeling, gathering my equipment from the basket, when I realised they had fallen quiet and in the silence I heard a chilling sound. A sliding of machined metal parts that ended with a clack. Someone right behind me had cocked a weapon and the sound had been gritty, coming from a dirty and uncared-for weapon, held in inexpert hands. I turned round very slowly, to face three policemen pointing automatic weapons at me, crouched in dramatic attitudes as if posing for a communist mural depicting the glories of the people's armed militia.

We were under arrest, accused of spying – having sneaked into the country in a ten-storey-high orange-and-yellow stealth balloon. A short while later our ground crew arrived and all of us were hauled off to the Moshi police station where, despite my properly stamped flight plans and a letter from the Director of Civil Aviation, we were questioned for the rest of the day. Our passports and cameras were impounded overnight, and after another five-hour grilling the following day we were finger-printed, had mugshots taken and were told we could go. Just one more thing – they wanted the film from the cameras. I threw a fit, but my desperate pleadings were ignored, they wanted all the film and, opening the cameras and exposing the film to the light, they took it.

We didn't start laughing until we were well out of Moshi. The moment I had seen those guns I had given Joan and the cameras a meaningful look, which she had instantly understood. 'Please, officer, my wife was hurt in the landing, can she just lie down in the car for a rest while I show you my documents?' Joan had unloaded our priceless film and replaced it with new, blank rolls. In due course I sent the police a signed picture of the balloon above the mountain, thanking them for their help.

The footage that Joan had rescued, and Ian's shots from the plane, would provide the climactic sequence for a film we were now making about ballooning in Kenya. The only way to get slow, silent, air-to-air shots of a balloon is from another balloon, so I bought an identical craft, took on a pilot and, gathering a group of friends to help, we spent several months having a wonderful time flying over the best wildlife areas in the country. We trawled for crayfish and sank in Lake Naivasha; had a lion chase and grab the trail rope as it snaked past her in the grass; landed where the cars could not retrieve us from thick forest near the Mara river and had to be rescued by an inflatable boat, in which we ran the gauntlet of several groups of hippos and surprised a cow elephant with her calf, who came deep into the river towards us and stuck her tusks through our boat, puncturing two of its five sections. It was a marvellous, totally unique and unrepeatable time.

One flight, at Amboseli, the national park at the foot of Kilimanjaro, turned into a hilarious ride with a group of Maasai warriors. They had helped us earlier with a sequence of retrieving the balloon from a lava field inaccessible to the recovery cars. Shouldering the rolled-up envelope like a long colourful snake, the men had walked it out to the car, while Joan carried their spears. I had promised to reward them with a ride, but on the chosen morning the wind was so strong that I could only get one man on board before the balloon was on its way. Fortunately we were carried across the flat, dried-up bed of Lake Amboseli, so Joan piled the warriors into the back

of a pickup and raced after us. Doing almost twenty miles an hour, there was no way that I could stop to collect passengers, so when Joan caught up I brought the basket down alongside the car and a warrior would leap in. The sudden addition of the extra weight made us sink, and drag and bounce along the dusty lake bed, while I kept the burner roaring to regain lift. With no vegetation, rocks or stumps to catch the bottom of the basket, this was the perfect surface for our madness. Again and again we repeated the manoeuvre until the basket was full, with the last warrior coming over the side head first, his naked legs bicycling madly in the air as we kangarooed along.

The Maasai threw themselves – quite literally – into this adventure that was so alien to anything they could have even imagined before. I have often taken them up in aircraft and helicopters, and have found that they are always ready for the adventure, but completely unmoved by the machinery or technology involved. I love their total indifference to the modern, mechanical world. Often, when flying across some dry and desolate plain, miles from water or any sign of habitation, I will see one or two tiny red-clad figures walking, with no destination in sight. Swooping low over them I know they will not look up or acknowledge the existence of me and my unimportant machine. They walk on regardless, complete men, and I am filled with admiration, and no little envy, for their self-contained serenity.

Whenever and wherever we flew our balloons, curious tourists would always approach us to ask if we were giving rides. When we finished filming I had two balloons plus all the backup equipment and a pilot, John 'Hawkeye' Hawkins, and I wondered if there were enough of those curious tourists to support a commercial balloon operation. Our neighbour, Jack Block, who owned Kenya's finest hotels and safari lodges, suggested that we start by stationing a balloon at his Keekorok Lodge in the Mara Game Reserve and, as he put it, suck it and see. At first

Hawkeye would have to go round the dinner tables persuading people that a one-hour dawn flight was a safe and wonderful way to see the animals, but Balloon Safaris very quickly caught on and became the first regular-departure balloon operation in the world. Thirty-five years later it is still going strong with many imitators.

Balloons weren't the only inflatables in our life at that time. Dr Grzimek still came out to Africa every year and for a couple of weeks I would take him on safari, filming whatever it was that captured his interest. One year he turned up with some inflatable plastic animals for conducting experiments in the wild. As an army vet he had done tests with horses to see how far a painting of a horse on a wall could be simplified and still be recognised. Apparently the horses would approach and nuzzle images as simple as a couple of connected rectangles – head and body – horse by Mondrian. Echoing those experiments, he now wanted to see how wild lions and rhinos would react to inflated models that had been specially made by a toy company in Frankfurt. When we blew them up we discovered they were badly made, garishly coloured, grotesquely obese and stank of plastic, so we did not expect great results. However, when we positioned the plastic male lion over the carcass of a zebra that had been killed by some lionesses it was obviously treated as the real thing. Several lionesses approached in a low, slow, submissive crawl to snatch food with a nervous growl from under its nose. Soon a male arrived and circled the dummy intently and, though it was considerably bigger than he was, he attacked from behind and gave it a classic crippling bite in the spine.

We thought rhinos would be less interesting, but not so. Our rubber rhino was a rotund fellow, some ten per cent bigger than nature, and as it kept rolling away in the wind I decided to hold it from behind and walk it up to a young male sleeping in open grassland. Rhinos have very poor eyesight and I got to within thirty yards before he saw me and rose to his feet. Making the tiny mewing sound that rhinos use to communicate, he ambled slowly towards me. I was ready to run, but my fear of being charged soon dissolved. This guy was not aggressive – he was coming, somewhat nervously, to greet me. He came right up and gently rubbed the side of his head against that of the dummy. It was extraordinary, all his clues pointed to an alien creature, greatly oversized and stinking of humans, plastic and the exhaust fumes we had used to inflate it. I could only surmise that despite all those wrong signals he was getting one very powerful piece of visual information which overrode all the others: 'Here is a very big rhino, I had better be friendly.' He moved to rub the other side of the dummy's head, then stood still as if waiting for my rhino to make the next move. I wasn't sure what that next move ought to be, so I signalled Joan to come forward in the car to extricate me while the rhino trotted off.

Bernhard decided he wanted some footage of himself in action, so we repeated the operation and the reaction was exactly the same, the rhino approaching, in peace as it were, and greeting the newcomer. We tried it with a third lone male and had the same result again, so I decided to approach a cow and her young calf I had spotted nearby, thinking their reaction could be even more interesting. At the time, I was clearly unable to differentiate between interesting and bloody stupid. From the calf's size I should have realised that the female was likely to be in oestrus, a time when calves are sometimes killed by rampant bulls, so male attention is certainly not welcomed by the cows. After the experiences with the solitary males I felt ridiculously secure as I walked towards her, hidden behind my plump bag of air. But this was no male I was approaching, and she had obviously decided the dummy *was* a male and she wasn't having any of it. Joan took a series of pictures – four per second – with a motor-drive Nikon that told the story in frightening detail. As I got to within about fifteen yards the cow came straight at us – me and the balloon – and hooked it with her long front horn three times in less than two seconds, ripping open the dummy's chin, chest and then belly as it flew into the air above her. Suddenly my cover was gone and I was on my own, just eight feet in front of an enraged rhino. By Joan's third frame I was gone from the picture. Had it not been for the rapidly collapsing dummy, deflating flatulently, which landed on top of the cow, I would never have got away. As I raced back to the car she spun round, trying to dislodge it, but I was gone. It produced some great pictures, but I don't think the experiment illustrated anything about animal behaviour – except perhaps why young men make up the most accident-prone cohort of the human race.

I walked, or rather sprinted, away from that accident, but shortly afterwards ran into another problem, this time while visiting Joy Adamson in Kenya's Meru National Park. Out on a walk nearby, Joan and I came across a four-foot puff adder on the trail, a fat, rough-scaled snake that bites a good number of people every year, mainly because they accidentally step on this beautifully camouflaged creature. Deciding that Joy would be interested in it, I pinned it down with a stick, got a firm grip on its neck and carried it back to camp. There, while Joy clicked away, I demonstrated its hollow, curved, three-quarter-inch fangs, perfect syringe needles, which oozed a drop of deadly honey-coloured poison from the tips as I pressed on the venom glands in the snake's large head. When the show was over I took it a short distance away and had just released it when Joy called out, 'Oh damn, I didn't have any film in my camera. Can you please do it again?' The snake was understandably pissed off with all the handling, and as I fumbled the catch it sank one of those long fangs deep into my right-hand index finger. The pain was intense and immediate. I knew this was an emergency.

Like most people living in the bush, Joy had some antivenin of indeterminate age in her battered and sweating fridge so, taking it with us, Joan and I headed straight for the airfield. We decided that flying myself was probably not a sensible idea so we asked a pilot who had arrived with some tourists if he would take us to Nairobi. Joan had put a pressure bandage round my upper arm to restrict the venom's flow into my bloodstream, but about half an hour into the flight I became nauseous and vomited. Joan gave me 10ccs of antivenin intra-muscularly in my arm and I lay back and tried to stay calm, knowing that a pounding heart would just distribute the venom even faster. The pilot had called ahead, so a taxi was waiting at the airport and a wheelchair at the hospital. When we arrived I was whisked inside, fainting and vomiting, and unable to inform them that I had already been given a shot of antivenin. Before Joan could pay off the taxi and follow us in, the young Asian duty nurse, assuming that my condition was due to the bite, gave me another 10ccs of antivenin – this time intrave-nously. Snake antivenom is made of horse serum, to which I was obviously allergic, as I went into anaphylactic shock – far more immediate and life-threatening than the bite. All I could do was keep saying that there was a coppery taste in my mouth, as if I were biting on a battery, then I started to go rigid, had trouble breathing and passed out. The next half-hour was touch and go as I was given powerful antihistamines and shock treat-ment to keep my heart going. I wasn't aware of all that was happening, but I do know what helped to save me: Dr Antonia Bagshaw, an excellent doctor with a deep, take-command voice that exuded confidence, had been summoned to handle this emergency. Through the haze I clearly heard her giving firm, calm orders, and I remember clinging on to that voice and knowing that with her in charge I would make it through.

When I came round in a private ward, Joan's face swam into focus, looking more worried than I had ever seen her. 'The doctors want to amputate your arm,' she croaked. Looking down at my hand, it seemed they might have a point. Puff-adder venom is essentially an anticoagulant: veins and arteries leak blood, causing severe local damage and internal haemorrhaging on a massive scale. I had been topped up with a large blood transfusion, and my hand was swollen to twice its normal size, and covered with large, blood-filled blisters. My swollen fingers curled round a blister the size of a tennis ball that filled my palm, and my armpit was coloured deep purple by subcutaneous blood. The normal prognosis with puff-adder venom is that it moves on through the lymphatic system, taking a few days before causing collapse and possible death. The hospital had recently treated a man who had been bitten as he fished out of the sea what he'd thought was a drowned puff adder. His hugely swollen arm was amputated immediately and he was out of hospital the next day. It seemed an extreme way to economise on hospital bed-nights, so Joan and I made it clear amputation was not an option. Joan had managed to contact Dr David Chapman in South Africa, the continent's snakebite expert, but because that country was still under apartheid rule, he would be refused entry to Kenya to treat me. Dr Louis Leakey, Kenya's famed archaeologist, offered to help and with his considerable influence was able to pull some strings. Chapman phoned and advised us to keep the arm packed in ice until he arrived, when he delivered his spot-on prognosis. I would keep the arm and hand, but the index finger would have to be amputated, and the crushed and damaged nerves meant that I would lose a lot of mobility and use of my fingers. He was correct and since then I have managed to break two of those remaining fingers, so my hand is now an awkward, arthritic claw, but it's still a lot better than a stump. The finger was presented to me in a jar of spirit with a proper museum label attached showing that it had been collected by *Bitis aerietans* – the puff adder.

The hot-air balloon gave us a unique, low and slow vehicle

for our filming, but I discovered another interesting form of transport when I read in an old newspaper that the German River Police had just bought a number of amphibious cars to patrol the Rhine. Called the Amphicar, it was a British Triumph TR2 sports car with a sealed underside and twin nylon propellers in protective housings at the rear. If it was good enough for the German police, I reckoned it was good enough for Africa, and I sent off the article to Dr Grzimek, recommending that we acquire one. Taking an Amphicar on the Nile would make a good programme, and it could be a useful anti-poaching tool in Uganda's National Parks where waterborne poachers were a problem. The Amphicar company liked the idea of the TV coverage they would get, so before long a bright-red convertible arrived and we took it out on to the Nairobi dam for some trials.

With the doors double locked I could race down the beach at forty miles an hour straight into the water, move a lever forward to engage the props and she would surge forward at a respectable speed. Coming out I put her into first gear so both prop and wheels were spinning as we reached dry land, pushing up even quite steep and slippery banks. Within days we were off to Uganda, a well-watered country where travel was constantly blighted by slow, rickety ferries that often broke down, forcing traffic to queue for hours. We had great fun driving past the lines of vehicles waiting to board the ferry, swerving and hooting as if our brakes had failed, then diving into the river with a huge wave coming over the bonnet. There would be yells of concern, followed by roars of applause from the waiting drivers and foot passengers, and we would circle and give the crowd a slow float-past of the vehicle that must have been every Ugandan driver's dream.

The Amphicar was a perfect machine for the Nile. We camped under a huge tamarind tree on a sandbank below the Murchison Falls, where a clear stream ran into the river. Each morning we would load the film gear on to the back seat of the car and drive straight into the Nile. Being broad and heavy, the car was a steady platform for filming from the water and if we wanted to get a different angle from the land we simply drove up the bank to a new location.

Murchison Falls is one of the most violent stretches of water on earth; where the mighty Nile, some two hundred yards wide, is squeezed through a twenty-seven-foot-wide chute that drops a couple of hundred feet. The roar of the confined water as it thunders through the gap, the flung spray and multiple rainbows provide a magnificent and intimidating experience. I decided that it would make a great picture if I were to get the Amphicar as close to the bottom of the falls as possible. Just below the falls the fiercely turbulent water circles in a powerful whirlpool, which I would need to avoid. (We once saw a dead hippo here – blown up with the gases of decomposition and as buoyant as a giant ping-pong ball – get sucked down into this funnel and not pop up to the surface again for a hundred yards.) The best position for photography was midstream about a hundred yards down from the falls, so I deposited Joan on a smallish rock there and headed the car towards the falls. To make headway against the current I stayed close to the bank, but was unexpectedly caught in a fierce backwash that swept me up the south bank almost to the foot of the falls, then swung the car out towards the whirlpool. Powerless to fight the current, suddenly I was whooshing round the rim of a deep vortex, looking down into its dark sucking mouth. The nose of the car dropped downwards into the pit and I scrambled to sit up on the back of the seat and almost leaped out, so sure was I that the car was going down. Joan, ready with her telephoto on the little rock, completely lost sight of the car and in her distress put the camera down. The steeply angled car made two or three giddying spins pointing down into that swirling maw, then we were violently flung sideways and out of danger.

All I could think of on that terrifying ride was that when I went down Joan would eventually have to swim for it, and be carried by the current that would take her down past the big sand spit where at least sixty large crocodiles waited for the river to deliver gifts such as this: and that I was going to die much more easily than she would. Slowly, shakily, I motored down to where Joan waited, ashen-faced, on her little rock. She had got a couple of shots of the car on its way upstream, but when I picked her up we didn't talk about the pictures, or what those few shots might have cost us.

Later, we moved further south in Uganda to the more placid waters of the Queen Elizabeth Park, where we stayed with the warden, Ian Ross, and his wife Sarah, and were joined by Dr Grzimek. He and I set off, rather late in the day, in a chartered plane to fly to the Kidepo National Park in northern Uganda, where we intended to remain for a week. Our pilot managed to get completely lost and just as it was getting dark we were lucky enough to see an old airstrip and make a safe landing. But this was a military field and it was in the Sudan. We were under arrest, at a dot on the map called Ikitos.

We spent the first two nights sleeping on the floor of the guardroom plagued by fleas and mosquitoes, until our captors told us that they could not decide our fate so we would need to be taken higher up the chain of command. Over the next two days we were moved in the back of a bouncing truck to Juba, on the Nile, where we were allowed to stay in the local hotel under heavy guard. We were sampling the local beer when in muscled a huge fellow in tiny shorts, looking like the white hunter in *Jurassic Park*. He bellied up to the bar and called for a beer in the thickest Afrikaans accent I had heard since Ace DuPreez had told me all about hippos.

'So what the hell are you guys doing here – you must have a blerry good reason?'

'We're all under arrest,' Grzimek explained.

'Oh ya, I saw all your guards, they asleep outside.'

He was a South African RAF flight instructor seconded to the Sudan air force to train their pilots on their new Fokker aircraft, or 'trying to teach these fokkers to fly Fokkers' as he so delicately put it in a voice audible in Cairo. We told him that the army brass in Juba couldn't make a decision about our fate, so we were going to be trucked off again, this time to a higher authority in Khartoum.

'Yurra man, that'll blerry kill you. Tell you what, I'm sure you've heard that "'N boer maak 'n plan" – us boetjies always make a plan. I'll arrange a training flight up there for tomorrow and you can come. No worry, I'll fix it.'

Fix it he did and he slept soundly in the co-pilot's seat as one of his students, the epitome of a wannabe pilot, complete with RayBans and chamois gloves, took us on an admirably smooth flight to Khartoum. That night we were still under armed and sleepy guard, but now in a grand old colonial hotel rapidly going to seed. The rooms were vast and high-ceilinged, with deep bathtubs, masses of hot, coffee-coloured water and – a touch I particularly liked after several days of squalor – enormous fluffy towels with a large blue monogrammed SR, for Sudan Railways, who ran the establishment.

The next day we scoured the town and found several of Grzimek's books, illustrated with my photographs. Presented with the evidence that we were just wildlife nuts, not photo-journalists aiming to speak ill of Sudanese democracy, we were told we were free to leave. However, as we had spent what little money we'd had left buying the books that set us free, we could not now afford airfares home. We turned instead to our diplomatic missions. With the help of the German embassy, Dr Grzimek was able to leave in a couple of days, but our pilot and I had to deal with the British High Commission, where a low-ranking official told us that we did not qualify as 'distressed

British citizens' and our case did not merit disturbing Sir John. To remind ourselves of the days when the British took a bit more care of their distressed citizens in hot places, the pilot and I walked down into town to see the statue of General Gordon.

Gordon had been a popular figure in Britain and in 1885, when he and Khartoum were under siege by fifty thousand Islamic rebels, a rescue party had been sent up the Nile to relieve the town, but as he went down under a hail of spears, Gordon didn't know that help was on its way. Feeling similarly abandoned, we gazed at Gordon seated on his bronze camel and were watching a herd of real camels blocking the traffic when we noticed a shiny black car flying a small union jack. 'Must be Sir John,' we said in unison and before it could work its way through the herd we caught up with the car and I opened the back door and asked, 'Sir John?' Those were days when an impulsive action like this was not met with fear and armed response.

'Yes, yes, get in, man, what can I do for you?' He listened attentively while we explained how his gatekeepers had been keeping us at arm's length. 'So sorry, tiresome fellows. Look, come home and Penelope will give you some tea while I sort this out. Absolutely ridiculous.'

Sir John was as good as his word, and the next day we booked out of our hotel and were on a plane headed home. I'd had no means of communicating with Joan, so she was unaware of all this, and was still staying with Ian and Sarah Ross. When I finally reached them I had a house present for our hostess – a couple of huge fluffy towels monogrammed in blue with her initials, SR.

6

The Bridge on the River Ntungwe

In 1962 we were doing some speculative filming in the Queen Elizabeth Park in Uganda, close to the Congo border, when the Ntungwe river rose so high it damaged the road bridge and left us marooned on its south bank. Five days later we were still stuck and running out of food, especially for the young parrot we'd rescued when it fell from its hole in a tree the storm had brought down. We were surprised when one morning the Uganda Parks' plane dived over our tent and dropped a message in a canvas bag. It was from the Warden, Frank Poppleton, and asked me to be at the Ntungwe bridge at noon for an important meeting.

In heavy rain we arrived at the bridge, which sagged drunkenly, the midsection still deep under fast-flowing water. On the other side a middle-aged gentleman in a new safari jacket was starting to wade out to the middle, so I hurried to meet him. The current was fierce and, hanging on to the barrier rails, we inched as close as we dared until we could hear each other over the roar of the water. What followed must be one of the strangest job interviews ever.

'Good morning, Alan, I am Aubrey Buxton of Anglia Television and we have just started a wildlife series called *Survival.*'

'Oh, nice to meet you, sir, so what are you filming here?'

'I'm not actually filming. I'm out here looking for subjects, but I'm also on the hunt for a cameraman for the series and I'd like to talk to you about it.'

'This bridge is going to be out for days, but I could drive into the Congo, round via Kabale and be with you tomorrow.'

'I'm afraid I have to leave this evening, but you come highly recommended by Frank. He tells me you are the man I'm looking for. Would you like the job?'

'I'd love it!'

'Splendid! Welcome to *Survival*. I'll write to you with the particulars. Now, as you are marooned, do you need anything? Food or petrol?'

Hungry, but not wanting to appear a wuss, I replied, 'We're fine, thanks, but our parrot is getting really hungry.'

A couple of hours later the plane flew over and dropped a large bag of parrot food, presenting me with the line that *Survival* had started out as it intended to continue – by giving me peanuts! This was the start of a wonderfully productive relationship with Aubrey and *Survival* that, with just one hiccup, lasted thirty-five years. I still owed allegiance to Dr Grzimek and worked with him for several months every year. We had plans in the pipeline, including a filming trip to Australia and New Guinea, so it was decided that from then on the German rights to anything I filmed would go to Grzimek and *Survival* would have the rest of the world. Everybody was happy.

Some time later, we had just got back from a trip filming the Karamojong tribe in northern Uganda, when I received a call from John Williams, the ornithologist at Nairobi's Coryndon Museum, with the startling news that perhaps as many as a million flamingos were nesting on Lake Magadi. Since the Grzimeks and I had loaded our sledge and failed to film them nesting on Lake Natron, the renowned ornithologist Leslie

Brown had also tried to wade out to the colonies and had almost died in the attempt, overcome by the heat and the corrosive soda that had got into his wellington boots and rubbed his feet and ankles raw. He spent six weeks confined in hospital, where he underwent extensive skin grafts. As the nesting behaviour of flamingos had never been closely observed or documented, this was a unique opportunity for us. All our camping gear was already in the trailer, as yet unpacked, so we stocked up on food and drove straight down to the lake.

At two thousand feet altitude, on the floor of the Great Rift Valley just sixty miles from Nairobi, Lake Magadi is like a forsaken corner of hell. With no outlet, the mineral-rich water flowing into the lake from the Rift Valley escarpment has evaporated for millennia, creating a toxic yet fertile brew of dark purple soup capped by a crust of sodium salts which covers most of the lake. Like Lake Natron, this glaring white crust, up to ten feet thick in places, reaches temperatures of up to 165°F (74°C) and blows off the lake in blinding, choking clouds of acrid powder. The El Niño year of excessive rainfall and flooding had raised the levels of all East Africa's lakes, and the increased production of spirolina, the blue-green algae that is the flamingo's main food, had fast-tracked the birds into breeding condition.

The flamingo is an extraordinarily specialised feeder. With long legs so it can feed in deep water, and a long neck to reach down to the surface when standing in the shallows, its design is improved further with a head and beak made to operate upside down. To feed it hangs its head down and submerges its bill a few centimetres. The fleshy tongue works like a pump to draw water into the beak, then force it out through rows of fine hairs, which filter the algae and push it down the bird's throat. In a year of algal bloom the weight of spirolina produced could be in the tens of thousands of tons. However, the water level of Lake Natron, where they normally bred, had risen so much that the flamingos' traditional nesting areas had been flooded and the birds were forced to move to Lake Magadi. Magadi is almost a carbon copy of Natron, except that the lake's surface is mined by Kenya's biggest commercial operation, the Magadi Soda Company, which uses giant floating diggers to chew away at the crust, mix it to a slurry and then pump it ashore to a grim, sprawling factory that produces soda ash for use in glass manufacturing. There are hundreds of employees, causeways out on to the crust, noisy trucks and trains operating, and constant human activity. But such was the flamingos' drive to breed that when they discovered the apparently ideal conditions at Magadi the birds ignored all the bustle and disturbance, and built their mounds of soda right in front of the factory.

Next morning, in the few cool minutes at sunrise, we walked out on to the crust and into another world. The stench and noise of any large nesting colony is always overpowering. Coupled with the intense heat thrown up from the soda surface, and the vivid pinks and crimsons of a million birds in glorious breeding plumage, the effect was one of complete sensory overload. With all the wild honkings and trumpetings, and the strange sight of the incubating birds, each sitting on its own six-inch-high cone of soda, with its long scarlet legs folded under it, and the tenderness of those clumsy, upside-down beaks dribbling liquid food to its single tiny chick, there was so much to record and all of it for the first time. The birds were constantly scraping up more mushy soda to add to their nest mounds, and reaching out with long necks to bicker with neighbours and steal bits of nest lining and feathers.

Black-and-white Egyptian vultures paced among the nests, looking for abandoned eggs. They normally have a bare yellow face, but eating large quantities of blood-red flamingo yolks had affected their pigmentation and their faces were now bright orange. On finding an egg a vulture would lift it high in its beak and throw it down hard repeatedly until it broke. Early in the century there had been reports that these vultures broke ostrich eggs by throwing stones at them, and it was easy to see how that behaviour could have developed. With egg throwing already part of their behavioural inventory, it was a small step – when presented with the super-stimulus of a huge egg they could not lift – to frustratedly throwing a rock instead – a few successful outcomes and the behaviour would have been cemented in place. There were marabou storks here too, over three feet high with the dress and demeanour of undertakers, stabbing their eighteen-inch pickaxe bills at a sitting bird until it was forced to stand and expose the tiny chick. A quick snatch and the chick would be tossed into the air and down the marabou's scabby throat. We later estimated that the storks, vultures and eagles accounted for the loss of some four thousand eggs and chicks every week. At first hyenas also came out to raid the colony and wreaked great damage, but the soda apparently hurt their pads, for their visits stopped after a couple of days.

The greatest cause of mortality among the chicks was to come as a complete surprise to us, however, and at a time when the young seemed well established. Eight days after hatching, the fluffy grey chicks climbed down off their nests on their short pink legs and started to band together. These groups coalesced until there were herds of several thousand chicks of different ages. Before long there were two herds of around three hundred thousand birds each, constantly on the move; flowing over the crust like pulsating grey carpets, chirping and squeaking, the pattering of almost a million tiny feet producing a surprisingly loud murmur. Except for a few thousand that stayed behind, all the adults would fly off in the morning to Lake Natron to feed. When they returned in the afternoon there was tremendous excitement among the chicks and we watched – almost in disbelief – as the adults walked around in the seething mob calling constantly until they found their frantically peeping chick. All this trampling by millions of feet turned the slushy surface into a saturated solution of soda about three inches deep. The surface crystallised in the intense heat and as the chicks waded through the pools they collected crystals on their legs, which within three days grew into apple-sized anklets on one or both legs. Death soon followed as a result of exhaustion or drowning.

We first noticed this phenomenon in a single dead chick and thought it a freakish occurrence, but within a few days there were thousands shackled, and after a week tens of thousands. Although the problem was on an epic, seemingly unmanageable, scale, we felt we had to do something to help. After trying several different ways of removing the anklets I discovered that, like a bangle, it was loose and not attached to the leg. Placed on to a hard surface and tapped with a hammer it shattered without hurting the leg. Trying to help would be a mammoth operation and we needed assistants. We soon recruited a number of volunteers, including a platoon of British soldiers who set up a camp for all the helpers. There was also a teachers' strike on at the time, so we were joined by an enthusiastic bunch of Magadi schoolchildren. On the factory scrapheap we found plate-sized metal discs which, placed on top of a nest, made a perfect anvil. With the soldiers and volunteers armed with hammers, and the schoolchildren bringing in armfuls of chicks, in a week of hard, hot labour we were able to free some twenty-seven thousand chicks from their shackles. Perhaps just as many died – they were spread over such a big area that we could not round them all up – but by keeping the main congregations of chicks away from the worst lagoons we calculated we had saved another hundred thousand from becoming shackled. Everything about the event was extraordinary, the phenomenal rains that set things in motion, the sheer size of the colony, which has never since been repeated, the great numbers of chicks afflicted with the anklets, and the size and scope of the rescue operation. Extraordinary, too, was the fact that these events, which had never been seen before – because they had always taken place in the most remote and unseen place – had this time occurred in full sight of a huge factory staffed by hundreds of workers. Since then we have seen that the anklet phenomenon occurs at almost every breeding site, but the chicks normally trek to the nearest freshwater springs, where the bangles dissolve before growing too large, so the mortality is much lower.

Since the breeding activities of the flamingos had been so well hidden from the Maasai living around Lake Natron, they had built up a wonderful belief about the birds. The nesting took place way out on the crust, invisible in the shimmering haze. Then one day great flocks of chicks would suddenly wade in from out of the mirage and arrive at the springs where the Maasai watered their herds. The chicks were greyish white, the colour of the soda, and often their downy coats were encrusted with crystals so the Maasai believed that the chicks were created by God out of pure soda, somewhere out in the middle of the lake. We took several groups of Maasai out to the Magadi colony where they were amazed to see that, like every other bird, flamingos hatched from eggs.

Before too long Joan and I were back on the Serengeti, filming a pack of wild dogs and their pups. Their burrows were behind Lemuta Hill, many miles from any road, and we camped near them in the Land Rover for several weeks, seeing no other humans except an old friend, Yank Evans – the man who had brought me the baby bongo. While Yank was staying we climbed up on to a kopje, an outcrop of huge granite boulders piled up on the edge of the plains, and were resting after an excellent picnic lunch. We were spread out like lizards in the sun, waiting for the evening light to start filming again, when an owl flew out from behind a rock ledge nearby, and Yank and I decided to investigate to see if it had a nest. Walking to the edge of the granite shelf and looking down, we saw, about eight feet below us, the body of a jackal, freshly killed. A dead animal doesn't flush a sleeping owl. The jackal had obviously just been brought there, almost certainly by a leopard that had probably moved away at the sound of our voices. It was nowhere to be seen, so I jumped down, not realising that the ledge we were on had a big overhang and that the leopard was tucked underneath, hiding. I landed with my knees bent and my backside stuck out right in the leopard's face. Growling explosively, more in shock than anger, she bit me twice in my left buttock. I spun round as she reared up at me and, locking my hands together, I clubbed her violently away. Fortunately it was a female, probably weighing no more than eighty pounds, and raking a claw along my right arm she tumbled down the steep slope and ran off.

I looked up to see Yank poised above me in a position I thought only occurred in Tom and Jerry cartoons. He had been about to jump down to join me and was now leaning out at an impossible angle, windmilling his arms frantically in an effort to pull himself back to the vertical. In a fine example of what's possible when pumped with adrenalin, he made it and within seconds I was also back up there with him. The bites were deep and had only punctured the muscle, not torn chunks out, but they bled extravagantly. A short while later I limped into the office of my Serengeti friend, mentor and Chief Park Warden Myles Turner, to report the event. Despite the blood trail across his floor he could see that I was not seriously hurt and I knew he would see the funny side of the matter.

'So what happened to you?' he asked.

'One of your leopards just bit me in the bum,' I replied, 'and I think I might sue you.'

He looked at me with his grey sniper's eyes. 'You know you're not allowed to feed animals in a national park,' he said and grinned. 'Now get out of here before I lock you up.'

*

143

Our next adventure was certainly our biggest challenge so far. In late 1962 Dr Grzimek had asked us to visit the Virunga volcanoes in the Congo to film mountain gorillas. Joan and I had seen these animals briefly on the Uganda side of the border, near Kisoro, where a short, rotund and lederhosen-clad German, Walter Baumgartel, ran a little collection of mud-and-thatch huts called the Traveller's Rest. Walter was a great character and host who loved to join his guests at dinner and tell stories – mainly, we thought, to distract them from the rather dodgy food. He had an African guide, a slim Bahutu called Reuben Rwanzagiri, and between them these two unlikely partners had invented the gorilla trek, which is now one of the main pillars of the economy of neighbouring Rwanda and a million-dollar business in Uganda.

In previous years, Joan and I had made a couple of trips with our safari clients to see the gorillas with Reuben. Our contacts had always been brief and noisy – some dark shapes in the dense undergrowth accompanied by ferocious screams as the dominant male, the silverback, warned us not to come any closer. Across the border in the Congo there would be no visitors, more open vegetation and more gorillas, some of which had had contact with George Schaller, the fine American biologist who a couple of years earlier had spent time there studying them. But while George was up in the mountains, the Congo had gained independence, the army had rioted and thousands of whites had fled the country. George had been compelled to cut short his research trip and we were unsure how safe it was going to be for us.

The Congo was a post-coup shambles of roadblocks and drunken soldiers, with the occasional swollen body buzzing with flies in a roadside ditch, and endless passes and permits required by surly officials of one sort or another. We stayed in the National Parks' guest house, a once handsome building that was now filthy and stank of beer, piss and vomit. Why is it that the first thing revolutionaries do when they take over is ruin many of the things they have fought for? We gritted our teeth for a few days while I tried to stay cool and be nice to the officious thugs we were dealing with – immigration, customs, police and military – all with their hands out, but not in welcome. However, as Grzimek had been helping the National Parks with vehicles and radios through the Frankfurt Zoological Society, the director was welcoming and helpful, and soon we were ready to head up to the gorillas.

We took on Senkwekwe, an old colonial-era park guard who had also worked with George Schaller, and assembled a troupe of thirty-five porters at Kibumba, a village at the foot of Mikeno, the fourteen-thousand-five-hundred-foot peak that is the most dramatic of the Congo's volcanoes. We set off up the steep and slippery trail to the small cabin at Kabara some five thousand feet above us. At one point the steepening trail led along the edge of a precipitous gorge several hundred feet deep called Kanyamagufa, roughly meaning 'where animals die'. A mournful howling came from the depths of the canyon and Senkwekwe said it must be a poacher's dog, tied up and unhappy because it couldn't get at their meat. After five and a half hours of climbing we arrived at Kabara and came out of the forest on to an exquisite little meadow. We paid off the porters, who scampered back down the trail as if they had just had a short stroll.

We took a look at the small, two-roomed cabin made of rough timber, that was to be our home for the next few months. One outer wall had been burned down by poachers, so the hut was freezing cold – icy winds tore in through the open side and rain through the rusting iron roof – but it was set in one of the most beautiful places on earth. The meadow was mown short by buffalo and surrounded by hagenias, one of Africa's loveliest trees. Long garlands of silvery beard moss, glistening with moisture, hung from every branch. At one end of the meadow was a shallow, crystal-clear pool from which a couple of black ducks sprang into the air and disappeared, quacking urgently, down into the valley. On one side the massive cliffs of Mikeno towered above us, wreathed in shifting wisps of cloud. On the other side the land sloped ever upwards towards the highest of the volcanoes, which often had a dusting of hail on its rounded peak, giving the mountain its name – Karisimbi, from *nsimbi*, meaning the small, white cowrie shells that were once Africa's main currency.

Just behind the hut was the grave of Carl Akeley, the American taxidermist who had first come to Kabara to collect a gorilla group for the American Museum of Natural History in New York, and had returned in 1926 with his wife to study the gorillas. Shortly after arriving he had contracted malaria and died. His greatest achievement had been to persuade the Belgians to declare these volcanoes Africa's first national park, and to extend it to include a huge area of plains, rivers, lakes and forest running north to the Rwenzoris, the fabled Mountains of the Moon. His legacy is arguably the finest reservoir of biodiversity on the continent. On arrival we found that his headstone had been defaced by poachers, presumably those who had trashed the cabin, so we restored it as best we could.

We were lucky to make contact with a group of gorillas on our second day. We kept our distance and after a while they settled down and started to feed. They were too far away to film but we were able to spend an hour watching them, which we felt was a great start. Over the next month we gradually managed to get closer and closer to several groups until we felt that we had gained their trust and could start filming them. Now that some mountain gorillas have been habituated in Rwanda and are visited by a regular flow of tourists, there are many thousands of people who will agree that meeting a gorilla – even when time constrained and in the company of strangers

– is a life-enhancing experience, a meeting that transcends the species' difference and which will be remembered for ever. Back then – when it was just Joan and me alone in the forest – forging an understanding with these creatures was a magical challenge.

George Schaller's meticulous observations were an invaluable guide, and his book, *The Mountain Gorilla*, was our bible, but we actually found working with the animals surprisingly straightforward. Compared to tracking on the savannah, following gorillas was simple. Frequent rain meant the ground was soft and easily imprinted by the knuckles of these heavy animals; the vegetation below the spreading hagenia trees – giant stinging nettles, wild celery and thistles – was easily broken or flattened by the passage of their large bodies; and the size and number of nests they built each evening, and where they spent the night, gave a clear indication of the make-up of the group. Unwilling to get up in the freezing night air, gorillas defecate in their nests and even the infants mark their presence with little droppings alongside the mother's. Every morning, once we had found their night nests, we would follow quickly at first, then slow down around ten, knowing that by mid-morning the animals would stop for a rest. This was when they made their day beds, bending down foliage to create a mattress that kept them off the wet ground. Now the adults relaxed and the youngsters would play, and it was the best time for us to make contact. We knew we were getting close when the droppings we found were still steaming in the cold air, and we could hear breaking vegetation, soft grunts and the sound of chewing. Then would come the defining moment when we were first sighted by the silverback, the alpha male of the group, who usually greeted us with a blood-curdling, screaming roar, an incredibly violent intrusion into the hushed symphony of the forest. We would catch his smell, always powerful but now amplified by his tension, a virulent addition to the forest's ambient fragrance of crushed herbs and wild celery. It's hard to describe it as anything other than an acrid bouquet of great depth – of sweat with a hint of smoke layered with intense overtones of urine, balsam and juniper that remain on the nose. Only a wine taster could adequately describe that pungent pong, but any pilots among you would recognise it as a weapons' grade version of the aroma that accompanies a trainee making his first solo landing.

The first time I managed to get a really good look at an alpha-male gorilla I was struck by an impression of enormous strength, but perhaps more by his great dignity. He looked as if he just *knew* he was magnificent, with his great helmeted head, massive shaggy forearms and jet-black coat with a saddle of pure silver. Every now and then he would emphasise his stature with a chest-beating display. This starts as a series of high-pitched whimpers and ends with a rapid whinny as he rears up on to his hind legs and beats his chest, which does not produce a deep drumming noise, as old movies would have us believe, but a flat pokka-pok slapping noise – actually a faintly ridiculous soundtrack to such an impressive display. Surrounded by his females, some of them holding infants with wide-eyed, surprised expressions and radical hair, he evoked feelings of family and kinship rather than the fear that a hundred years of bad press had prescribed.

We concentrated on two groups that usually nested within a few miles of our cabin. After a couple of weeks they began to accept our presence and we would quietly move into a position where they could see us but not feel threatened, and there we would stay, watching them. Gradually they relaxed, allowing us to creep closer, always staying in their sight so they did not feel we were stalking them. Finally the adults began to ignore us, reverting to their normal behaviour, and we felt we could start filming. But getting the cooperation of the gorillas was only part of the problem. Trying to film jet-black animals in overcast forests was difficult and often impossible. Fortunately the gorillas enjoyed the warmth of the sun as much as we did and would often make their day beds in open glades where they could sunbathe for an hour or two before heading back into the gloom. Here at last we were able to get film of them making their beds, feeding, grooming each other and chest-beating; and the youngsters wrestling and swinging on vines. In the twenties Carl Akeley had brought his movie camera here and managed to get some shots, and a few decades later George Schaller had taken still photos, but we were recording gorilla behaviour and habits never filmed before and every wet, cold day was packed with effort and reward.

This was a tough assignment. At high altitude the air is thin and hard exercise is doubly tiring. The trails led up and down steep ravines and often through beds of eight-foot-high stinging nettles, so dense that our clothes became sticky with the toxin that dripped from the nettles' long spicules. The pain of this was the one thing that really distressed Joan. She wore gloves, and cutting up an old kitbag I sewed heavy canvas patches on to her sleeves and trouser legs to give her some protection. It was painful just to watch a gorilla hold a stinging nettle stalk between finger and thumb, strip all the leaves, and chew them with no sign of discomfort.

Only Senkwekwe had stayed up on the mountain with us, to help with all our camp tasks. He was the sole African among the scouts and porters who enjoyed the contacts with gorillas and was happy to stay up in the high, cold forests. Totally indefatigable, and immune to cold and rain, he had a wonderfully debonair, gallic insouciance that he must have picked up from some former Belgian employer. It rained almost every other day – the sort of days that you wished, just once, could end with a long hot bath. Instead, we walked to the pool near our cabin and threw icy water over each other before climbing into as many layers of clothing as we could and sitting round a smoky fire. It was wonderful!

Every day we found evidence of poachers in the area. Tracks of men and dogs, old camps, snares and butchered buffalo or duiker antelope. One day we saw smoke coming from the trees down on the Bishitsi flats, a large, fairly level area overlooked from the track that ran round the slopes of Mikeno. Senkwekwe and I left our packs on the trail and headed down the thousand-foot scarp to investigate while Joan walked back to the cabin. Down on the flats we found a camp that must have been in use for a year or more, judging by the number of trees that had been cut for firewood. It was empty now, but had clearly been occupied recently and obviously the poachers were completely undisturbed up here. We kept going towards the smoke and started to hear dogs, men shouting and the distressed screams of an elephant. It was getting late so we sped up and suddenly came upon a primeval and barbaric scene.

A very distressed elephant stood in a wallow, throwing mud and water over itself to cool off. It had several spears sticking out of its body and had obviously been hounded to exhaustion. A pack of dogs raced around, yipping and harrying the animal, and we could hear the men, who were keeping a safe distance away and had lit a fire. We guessed they were going to wait until morning, by which time the elephant would either have died or be much easier to finish off. Some of the dogs had found us and I felt sure they would give us away to the poachers who were present in quite big numbers. The light was fading rapidly and Senkwekwe and I decided that before we were discovered the best thing to do would be to shoot the elephant. This would put it out of its pain and at the same time scare off the poachers, who would assume we were a ranger patrol. Senkwekwe, however, had no idea how to shoot an elephant, so he handed me his rifle and three rounds, which were all he had. The moment I felt them in my hand I knew something was wrong. The bullets were of two different calibres, both of them wrong for this rifle – there was nothing I could do. The dogs were now concentrating more on us than the elephant, barking excitedly and being quite aggressive, but it was almost dark so we were able to move quietly away without being seen.

Now all we had to do was find our way back to camp, about three miles off and a thousand feet above us. We took a straight line up the Bishitsi escarpment, stumbling through steep nettlefields and crossing ravines, heading for the dark outline of Mikeno against the brilliant star-filled sky. Seeing something glowing on the ground, I found a large phosphorescent fungus, which I crushed over Senkwekwe's back, a faint but valuable beacon to follow in the deep of night. Then we saw a tiny light way above us; a worried Joan had come back out to the head of the Mikeno track and hung a lamp on a tree to guide us in. Four hours later we collected it and our hidden packs, and cruised the last mile, finally reaching the cabin at midnight. Early next morning the tireless Senkwekwe ran down to the park headquarters and arranged for a patrol to head for Bishitsi. By the time they got there the elephant was dead: the tusks, much of the meat and the poachers all long gone.

My most memorable contact with the gorillas came one day towards the end of our three-month stay, when I was out alone looking for a way to climb the towering peak of Mikeno. I have never felt the urge to conquer mountains 'because they are there', but every day I had looked up at that dramatic peak and finally felt that I should at least investigate the higher slopes. But one close look at those vertiginous and slippery walls and I decided it was not for me. I was heading back to camp in a heavy downpour when I lost my footing on a steep stretch and slid down the trail on my back. I flew for some distance and when I stopped I was within fifteen feet of a whole gorilla group. Although no alarm calls came from them, I lay there tense, frightened and expecting attack, for I had violently invaded their space. Over the weeks we had learned their flight distance – the line beyond which an approach made them uneasy and caused them to flee. I had skidded to well beyond that line and up to the closer one which, if overstepped, was likely to trigger an attack.

The group were familiar with me, which certainly helped, but it was the freezing rain that saved me. The females were huddled under a sloping hagenia, pressed together with their infants held tight against their bellies. They muttered, shifted nervously and looked towards the silverback for guidance. I was closer to him than the others, but fortunately I had not split the group. He was sitting out of the rain below a massive horizontal branch, with his great arms wrapped round his upper chest and neck like a thick hairy scarf. Some raindrops sparkled on his dark coat, but he was basically dry. He shifted his arms a little, looked at me with his calm hazel eyes and I started

breathing again, knowing somehow that with our almost identical genetic make-up, he and I understood each other.

With creatures that are so close to us, that essentially inhabit a similar perceptual world to us, it is easy to put human thoughts into their minds. I was suddenly deeply aware of the understanding that I felt pass between us in those few tense seconds and I had no hesitation in anthropomorphising. I had been incredibly rude intruding on his group and had given them a bad fright, but he knew me, and knew I was harmless, so he was damned if he was going to run around in the rain, wave his arms, kick up a fuss and get soaked. Still, it would be wise for me to move on. We exchanged a look that seemed to confirm my interpretation and, pulling on exposed roots and stumps, I slid my way another thirty yards down the trail before slowly standing up. I gazed back at that peaceful group with a powerful surge of emotion and gratitude for the rare experience they had given me. The silverback was still looking at me. 'Thank you,' I whispered and would like to think he gave a little nod of recognition. I would never better those few precious moments. It was time to think about leaving.

A few days later we declared a rest day, cleaning cameras and doing our washing in preparation for our departure, when we heard voices coming up the track and about twenty porters appeared followed by a white couple. Outraged that someone could just turn up unannounced in our private world, I gave them a very cool welcome, suggesting that they camp at the far end of the meadow. The parks director had promised we would not be disturbed and had laughed at the thought that any tourists might turn up in his benighted country. Yet here was John Alexander, a Kenya safari guide, accompanied by a tall, dark-haired, intense young woman who limped in on a heavily bandaged ankle. She had apparently cracked an ankle bone earlier in the safari, but had insisted on climbing to the saddle and was absolutely determined to see gorillas, so while we did not welcome the intrusion we were certainly impressed with her spirit.

For the next three days Alexander tried to show her gorillas without success. His guide located them each day, but Alexander carried a holstered .38 revolver, firecrackers, and travelled with two armed guards so they never came close enough to see them. On their penultimate day the woman invited us to a late tea outside her tent, which she had decorated with wildflowers and furnished with cushions of moss. Tearfully she told us how hard she had worked to make this trip, how disappointed she was with Alexander and how tomorrow was her last chance to realise her dream. She told us about her work with crippled children as an occupational therapist back in Kentucky, how she had saved for years to be able to come and producing some blackberries she had picked in the forest she pleaded with us to take her along the next day. The sun was going down, sending shafts of golden light through the beard moss hanging from the hagenias, it was no evening to be uncharitable; we knew a group were not far away, so we agreed to her request. The next morning the gorillas gave her a command performance. The young played and suckled, the adults climbed and broke branches, the silver-back screamed and beat his chest. The woman was overwhelmed by the experience, lavish in her gratitude and said she was sure that one day she would come back. She did and became a life-long friend. Her name was Dian Fossey.

Our plan, once we felt we had enough gorilla footage, was to climb and film Nyiragongo, a volcano almost unique in having a mile-wide lake of molten lava in its crater. The aluminium huts that had once stood at the base and near the summit had been dismantled and stolen, so we took a small tent, an inflatable mattress, sleeping bags and two days' supply of food. We were fit from our high-altitude exertions but the climb was steep and we planned to leave our porters at the site of the stolen cabins and make the last lap alone. On the way up we passed through a field of ten-foot-high chimneys of lichen-covered lava that the porters called 'the soldiers'. Eerily similar to the terracotta regiments uncovered in China, they stood in ranks looking out over vast black lava flows. Every few years an eruption bursts out of the flanks of Nyiragongo and lava pours down the slopes, enveloping the forest and covering earlier solidified flows in its path. The eruption that created the soldiers had produced particularly hot and fluid lava that had been described as flowing almost as fast as water. It rushed into the forest, sloshing high up the trunks of the trees, where it cooled and hardened. The whole forest burst into flames and disappeared, resulting in these ranks of tall hollow columns, perfect castings of what had once been trees.

Leaving our porters and hitching up our heavy rucksacks, we set out for the rim under lowering skies whose threatening grumbles matched the sounds that we could now hear coming from the crater. Soon it began to rain heavily, then to hail large stones that pounded us painfully non-stop for fifteen minutes as we slogged upwards. The storm ended just as we reached the rim of the volcano. It was getting late, but at that very moment shafts of evening light illuminated the most breathtaking natural scene I have witnessed anywhere in the world. The dark vertical walls of the crater drop down to ledges in several three-hundred-foot jumps and at the bottom, over a thousand feet below us, a vast cauldron of blue-black lava heaved and bubbled. All over the surface the crust ruptured in long livid gashes, ejecting fountains of orange lava high into the air, then healing, only to open somewhere else. The whole rim was pure white, crusted with several inches of hail, and as it melted scores of slender waterfalls cascaded down around the walls to turn to steam before they reached the lava. It was a scene from creation, an elemental battle between fire and ice that I longed to get on film.

I had the camera set up in seconds, but we were enveloped

in warm humid air and it was impossible to clear the condensation from the lenses. I tried everything but before I could even start filming the lenses had steamed up. To hell with it, I was not going to miss another second of this spectacle unsuccessfully struggling with equipment. So Joan and I sat with our feet hanging over the abyss and watched as the waterfalls shrank and one by one were extinguished. The sun slid lower, and as the floor of the crater moved into deep shadow the whole surface now glowed red, and the ragged purple and orange gashes became more vivid. Waves of heat rushed up towards us and we were blow-dried, then warmed and finally driven back as the icy covering on the rim disappeared before our eyes. Would that my cameras had worked so that I could have shared those moments with the world, but in my heart I know that no camera, or photographer, could have done justice to that apocalyptic scene. It was the last night of the assignment and it could not have been bettered. Warm on one side and cold on the other we lay out under an expanse of sky, dark and star-studded except for a column of flickering orange smoke that rose straight up out of the bowels of the earth, whose incessant dyspeptic growling soon lulled us into exhausted sleep.

On our last day, just as we were about to climb into the Land Rover near the National Park headquarters, a surly soldier came up and said we must wait for his officer who wanted a lift to Goma, the pretty little lakeside capital of Kivu Province. After a few minutes I asked where the officer was and was told that he was having his lunch, so I went into the café to look for him. Fat, shiny and dripping with medals, he was tucking into a huge plate of meat and rice and obviously in no hurry. I rather bluntly told him I had an appointment in Goma, so would he please eat up. His reply indicated that he was far more important than me, or anyone with whom I had appointments, and he was enjoying his meal. By this time I had already had a bellyful of the Congo,

158

and my veneer of diplomacy, insubstantial at the best of times, slipped away. I told him he had five minutes and walked out. As soon as she saw my angry face Joan knew I had blown it. A voice from inside summoned the surly soldier and when he came out, unslung his rifle and positioned himself in front of the car, I knew we were in for a long wait. In fact, fatty came out after about fifteen minutes and started to climb into the front seat. Foolishly aggravating an already tense situation, but young and newly married, I said no way, that is where my wife sits. The surly soldier said no, this is where the officer sits. Before I could growl myself into really deep trouble Joan climbed into the back, giving me a conspiratorial grin and saying, 'Please let the officer have the front seat.' I started to object . . . then I remembered.

Earlier that day we had found a large chameleon crossing the road and, wanting to film it later, had put it on the central rear-view mirror, where it could not climb down. Chameleons terrify most Africans, and as the officer was manoeuvring his belly into the front seat his face came level with and inches from this harmless reptile, which rolled its gun turret eyes at him and inflated its bright orange throat. In the next few sweaty and scrabbling seconds our warrior's attitude changed from an arrogant sense of entitlement to a panicked demonstration of what the Congolese army did so well – full and ignominious retreat. He blanched, groaned fearfully and flung himself out

of the door. Joan was already slipping out of the back in anticipation of a military withdrawal and both soldiers gratefully took her place, where they sat rigidly the whole way to Goma.

This was a typical encounter. It had been a tough assignment: physically tiring we could handle, but despite our spiritually uplifting encounters with the gorillas and the spectacle of the volcanoes, it had been an emotionally exhausting trip. The Congo was sliding into anarchy. Since Independence in 1960 various political factions had been fighting for control of the country's vast mineral wealth. The CIA had helped the thuggish Mobutu to seize power in a military coup, the Russians had backed a different thug, Gizenga, while various mining interests had supported their man Tshombe, who had led the breakaway of Katanga Province, with its rich mineral pickings. For three years the Congo had been a blood-soaked political football, as the West and the Soviets armed different factions, and the United Nations muddled along trying to stabilise the situation. White people were suspect and time and again I'd had to leave Joan on the mountain while I went all the way to Goma to renew permits or submit to questioning about our activities – accused of being spies, diamond buyers, or of trying to catch a baby gorilla. Since I was working with Frankfurt Zoological Society (who were paying the rangers' wages), the parks' authorities were supportive, but the army and police were a constant pain. Joan would worry until I returned, and I would worry that she was alone up there, no matter how unlikely it was that trouble would reach that far. A little later the cumulative effect of all that stress would come home to roost in Joan's body.

Back in Nairobi, Joan began to show signs of a strange malady. She felt weak and uncertain that her limbs would do as she wanted. Her eyelids began to droop and at times were completely closed. She was diagnosed as having myasthenia gravis, a rare auto-immune disease caused by a defect in the chemical connections between nerves and muscles. In the early drug-dart experiments we had done in the Serengeti we had used a concoction that broke down acetylcholine, the compound that acts as a neurotransmitter passing instructions from nerve to muscle, thus rendering the animal immobile. The same sort of thing was happening to Joan and it was terrifying. Suddenly this extremely fit young woman was walking warily, putting out a hand to steady herself and viewing the world anxiously through half-closed eyes as her eyelids drooped uncontrollably. Apparently this little understood chronic disease can be brought on by exhausting the system to the point where the body shuts itself down to avoid further damage. The extremes of physical exertion, gruelling conditions and the psychological stress of roadblocks and drunken soldiery had driven Joan's body to that desperate end point: and we were told there was no cure. Aristotle Onassis, the billionaire shipowner, was afflicted with it and ran his empire with his eyelids held up with sticky tape. If he couldn't find a cure what were our chances? Since then, thankfully, treatments have been developed that cure the disease or at least allow people to lead fairly normal lives, but the only advice we were given then was to suggest Joan take it easy. In fact, she could do little else, so I wound back our film projects and she embarked on a long period of rest.

At Naivasha Joan would sit on the veranda and look out on our garden – where we counted more bird species than had been recorded in the whole of Great Britain. She slowly mended but grew tired of the repeated visits to doctors who could only advise more rest, so she accompanied me on a short trip up to the Aberdares, where I was going to film chestnut winged starlings, which nest behind the waterfalls high on the moorlands, and ring them for a study. We had a comfortable camp overlooking the falls where Joan could sit and watch the duiker antelope that came scavenging around the tent and the clawless otters hunting for crabs in the shallows.

One afternoon I saw Joan had walked to the edge of a narrow tributary about a yard wide. She was obviously trying to steel herself to jump but feared she would not make it and was rocking back and forth, torn by the conflicting emotions. I was in tears watching a woman that I knew could run and jump and climb as well as any man, yet who stood frozen with terror at the prospect of a three-foot leap. Then I saw her brace her shoulders and fling herself at the other bank with a wild yell. She just made it, stumbled, then righted herself and spun towards me flushed with triumph and a big smile. I was running towards her as she shouted, 'This is the source of the Tana! [Kenya's biggest river.] Tell those ruddy quacks I don't need them any more. I just jumped the Tana!'

The jump had dislodged her glasses, so I pushed them up and we walked back with our arms round each other, and she sat outside the tent watching me set up a mist net behind the falls to capture the starlings as they came in to roost. The low sun was shining on the giant senecios with their mossy trunks and huge silvery cabbage leaves. A rainbow repeatedly appeared and dissolved as waves of misty spray rose from the falls, and I noticed that this mist was clinging to my net and weighing it down so much that the bottom third was now underwater. When I waded out in the icy water to retrieve the rig I discovered that it had netted a couple of fat trout, more than I had caught in days of casting into that pool. We grilled them right there on the riverbank and toasted Joan's progress with pure mountain water from the stream. She was on her way to recovery, but neither of us would ever get over the news that the disease had somehow accelerated her menopause, which could mean she might never be able to have children. We were both twenty-six.

7

Going Solo

In between all our adventures at home in Africa, we also spent six months filming marsupials in Australia and birds of paradise in New Guinea. After my childhood dreams of South America I had finally got to the Amazon and we had spent a very successful few months filming in the Galapagos. These expeditions were made for *Survival* and Dr Grzimek, for whom I had now made about a dozen thirty-minute films, and we left the Galapagos buoyant with plans for a film that could easily fill an hour. I had shot some unique sequences, with a particular story in mind, so I was looking forward to taking part in the editing process, something I had not been involved with on my earlier assignments.

When we arrived at *Survival*'s offices in London I was surprised to be told by Colin Willock, who produced and wrote all their films, that I was a good cameraman, but I should stick to that. He did not want me sitting in (he made it sound like butting in) on the edit and completion of the film. I put my case to the chairman, Aubrey Buxton, who had long been a great supporter of mine. I showed him my outline for the Galapagos story, but he felt he should back Colin's decision. I really wanted the experience of the next step in film-making and sadly concluded that if they wouldn't give me the opportunity maybe someone else would. When my request was again turned down I resigned, and took the train to Bristol and the BBC's Natural History Unit.

On the way I went over a few ideas that had been rattling around in my head for some time. I presented outlines for three fifty-minute films that I called 'Portraits'. The first was 'Portrait of a Lake', about Lake Naivasha which, apart from being a stunning setting with some good stories, was on our doorstep and would involve little travel and other overheads. The other two portraits would be of the crystal-clear Mzima Springs in the semi-desert country of Tsavo National Park, and the life in and around a giant baobab tree.

Almost all wildlife on TV at that time involved a personality who took the viewer on a trip – usually to a national park – looking at whatever he could find of interest. Many were really quite superficial travelogues with animals, and seldom examined anything in detail. I promised that my portraits would have ecological themes that had not been seen on television before. There would be underwater film of hippos and crocs, time-lapse and slow-motion photography, I would get inside birds' nests, look at the relationships between animals. The films would not need a personality or presenter to make them work; they would be driven by the unfolding of the natural story.

David Attenborough, who was then head of BBC2 and in a suit instead of his shorts, was, of course, a superb naturalist and had earlier made many films with himself as presenter, so he understood my aims, but raised his eyebrows a little at my claims. However, he agreed to take the UK rights to the three programmes, with the proviso that I should shoot some 'presenter' material just in case, as he doubted we could get away with pure wildlife. Beyond that he couldn't give me much money, but I would be free to shoot, edit and write the films as I saw fit.

I headed back to London to give the good news to Joan, who

was about to fly back to Kenya. I called Heathrow and just caught her. I told her excitedly about the deal, but when I mentioned what we would get per programme she insisted, 'We can't do it for that.' She had always done our accounts and I had little idea of financial realities – my idea of a business arrangement was what had been shouted across that flooding Ugandan river. 'You'll have to ask for more,' she said. Then, 'Hang on, guess who's here?'

When I phoned the airport they had paged Joan to take the call and her name had been heard by an old acquaintance. 'It's Dian Fossey! She is on my flight to Nairobi. Leakey has asked her to go up to study the gorillas!' Joan was obviously as gobsmacked as I was by this news. Leakey had sent Jane Goodall and her mum off to Lake Tanganyika, which had been a pretty tough assignment, but the Congo was an altogether different and much more dangerous undertaking. Shaking my head in disbelief, I wished them both a good flight, then rang the BBC and asked for David. I still had no idea what these productions would cost, but felt I could only ask for a little more. 'I thought I might hear back from you,' he said with a chuckle. 'All right, but they'd better be good.' Joan said we still couldn't do it – but all I cared about was that we were going solo.

Among the pile of mail that awaited us when we arrived home in Kenya was a letter from Dr Leakey. As I was familiar with the volcanoes, the mountain gorillas and had contacts in the area, would I escort Dian Fossey to the Congo and set her up in a study area, he asked. Despite my misgivings about the plan there was no one else who could do it. Joan spent a day helping Dian buy the sort of clothes and supplies that she would need up on the volcanoes and then I set off, with Dian following in a clapped-out old Land Rover that Leakey had provided. We had the usual bureaucratic delays at the border, then had to negotiate with several police or army roadblocks, but after only two days we had all the necessary clearances. I hired some porters and we headed up the familiar trail from Kibumba village. I had located Senkwekwe, the park guard who had helped us the first time, and I recruited a couple of camp staff, one of whom claimed to be a cook.

On our arrival at the beautiful Kabara meadow, we found that poachers had again been living in the wooden cabin and it was a complete shambles: half the roof was now missing, more timber walls had been used as firewood, and once again Carl Akeley's grave had been vandalised. We put up tents for Dian and her staff, dug pit latrines and collected a large pile of firewood before the porters disappeared back down the mountain. I could only stay one more day, so Dian and I got out early the next morning and soon crossed some two-day-old tracks of a gorilla group. There was probably not enough time to catch up with them, but it was an opportunity to give Dian a quick course in gorilla tracking, but I turned round to find she was gone. In her excitement she had taken off at high speed following the clearly defined trails through the soft vegetation – but she was headed the wrong way! I sat down and waited, and after ten minutes or so a flustered-looking Dian came back down the trail and in embarrassed silence sat down beside me. She later recorded that I said, 'Dian, if you are ever going to contact gorillas you must follow their tracks to where they are going, rather than backtrack trails to where they have been.' Apparently this was delivered 'with the utmost British tolerance and politeness'. Doesn't sound like me – I must have been pulling her leg.

Dian was a brave and determined woman, who has been too readily denigrated for her behaviour in later years without consideration of the awful realities of her existence. Going out into the forest every day, she started to establish contact with the gorilla groups, having frightening brushes with elephant and buffalo, and peeing herself when she was first charged by a silverback. She also suffered from emphysema, a great disadvantage at altitude, and she was terrified of heights, so every time the gorillas crossed a ravine on a fallen tree she would have to go the long way downstream to find a less arduous crossing. But she was settling in and making some good observations when, after six months at Kabara, in July 1967, she was ordered off the mountain. Fighting had broken out in nearby Bukavu, on Lake Kivu, between white mercenaries and the Congolese army. The routed army was taking its revenge, the radio called for all whites to be killed, and Dian was taken to a nearby army camp where she was locked up, humiliated and raped. In later years some claimed the rapes were a fabrication to burnish her legend, but Joan and I were her closest friends at the time and we knew it to be true. (Indeed, anyone aware of the behaviour of the Congo army, then and now, wouldn't doubt it.) At the time she kept quiet about it for fear that Leakey and her other supporters would not let her continue. She knew there were gorillas in Rwanda – comparatively stable and just across the border – and she was determined to find them and continue her study. In Rwanda she made friends with an adventurous woman, Alyette de Munck, who had lost her farm to rebels and had been forced to flee the Congo. Her husband had just died and together these two wounded women planned to search for the gorillas on the Rwandan volcanoes, Karisimbi and Visoke. Also arriving to help out were Alyette's son, her nephew and one of their friends. The boys had just graduated from college, bought an ex-army jeep, and were on their way from Nairobi. At Kisoro, the Ugandan village where the road branches to Rwanda and the Congo, they took the wrong turn. Arriving at the Congo border post they were arrested as mercenaries and taken to the same camp where Dian had herself been held captive, where they were tortured for days and put to death in God knows what awful fashion.

The women were shattered by grief, but they somehow found the strength to keep going. They found gorillas, built a camp and after a while Alyette left Dian to relaunch her study. I had been asked by *National Geographic* to film the project, but the puff-adder bite had put me out of action at the time, so my friend Bob Campbell took over the assignment and moved up to Dian's camp.

Over the years she came to hate the poachers whose wire snares killed the forest antelope and buffalo, and sometimes jammed tightly round a gorilla's wrist so that the hand rotted away. She also came to mistrust the park authorities, some of whom were involved in killing gorillas to capture the babies for sale and to sell hands and skulls to tourists. She employed a number of men to collect the snares and chase poachers from the area but, frustrated by the lack of help from the authorities, she would order her men to tie up captured poachers whom she would beat, sometimes with stinging nettles. She was certainly going too far, but she was virtually alone, her head and heart still full of awful memories, with no one to turn to for help.

Later university researchers joined her camp, and although these youngsters came into an established camp where Dian had spent years habituating the animals that made their studies possible, few of them were prepared to help with her anti-poaching initiatives. Some took the moral high ground and said what she was doing to stop the poaching was wrong, while others sympathised with her frustration but simply did not want to jeopardise their projects, they just wanted the data for their PhDs and to get out.

The last straw came when Digit was killed – Digit, whom she had first seen as a ball of black fluff, who was the first gorilla to reach out and touch her, and whom she had watched grow into a silverback with his own group. For Dian, Digit was a trusting, trusted friend – whom she found dead one morning from multiple spear wounds, his head and hands hacked off, his faith in humanity betrayed. Nothing in her pain-filled past could approach the agony of that morning. She had always wanted a family and now she was losing the only family she had in the most horrible way imaginable. Her drinking got worse, she behaved outrageously and even her defenders started to distance themselves. A couple of her disciples suggested that the way forward was to open some of the groups to visits by tourists, which would provide hard currency and jobs for those living near the mountains, and thereby provide a reason to protect gorillas. For some time this had been the rationale for conservation across East Africa, but Dian did not like the idea; she was exhausted and the Rwandan government wanted her out.

She went to America where she spent four years writing up her results, lecturing, writing her book *Gorillas in the Mist*, restoring her health and her desire to return. By now she was world famous so, in 1984, Rwanda welcomed her back and so, most movingly, did her gorillas. In her absence more animals had been killed and there were now poachers all over the park. Before long she went back to her reckless ways of dealing with them. Perhaps it was this, or her opposition to tourism – in which some powerful political figures had strong interests – that led to Dian being killed in her hut on the mountain by an unknown person with a machete. Since then the Mountain Gorilla Project, based on tourists visiting the gorillas, has grown into a wonderfully successful operation that is today one of Rwanda's main sources of foreign exchange and gives hope that these creatures may be saved. I have seen tourists burst into tears at their first sight of a gorilla and the trusting look in those calm brown eyes – such is the impact of these magnificent animals. For all her faults, Dian fought for those creatures with every fibre of her tormented mind, heart and body. It was she who put mountain gorillas firmly into the world's conscience: she was heroic and her legacy is huge.

All this was yet to come, of course. When I left the visibly frightened Dian up in the Congo volcanoes it seemed unlikely that she would go on to make such a mark. I drove straight back home to Naivasha, eager to get started on our project for the BBC, which we decided to kick off with a film about the beautiful lake that we considered our front garden. Then, on my arrival home, with a flourish and a 'dadaaa', Joan presented me with a heavy card, embossed with the royal coat of arms. We were invited to attend a Royal Performance of *The Enchanted Isles*, our Galapagos film, at the Royal Festival Hall in London. It was a charity event to raise funds to buy a new ship for the Darwin Research Station on the islands. Prince Philip had done the narration for our film, and we would be presented to him and the Queen. As we were no longer with *Survival* we would have to pay our own fares, accommodation, black tie rental etc., which would take a big bite out of our mini-budget. We decided that we should go in any case and the *Survival* team was good enough to send a limo for us on the night!

On the flight back to Kenya we had time at last to talk about our new films. I had realised that while Lake Naivasha would make a lovely subject, it did not have the potential for a real attention-grabbing debut. I wanted to start with a bang, with something that had never been filmed or even seen before. Joan worried about the extra costs of abandoning the plan to film close to home, but agreed that the biggest bang for our limited bucks would be to film at Mzima Springs. I wanted to spend at least a year camped near the springs, and while Joan started planning I designed and built a couple of pieces of equipment that we would need to film underwater in safety.

Mzima Springs, in the Tsavo National Park, have their source twenty miles away in the Chyulu Hills, a fifty-mile-long chain of several hundred ash cones rising up to around six thousand feet. Despite heavy rainfall and lush forests of olive, fig and juniper, there are no waterholes on the Chyulus, and no rivers flow down to the parched red bush country below. Almost all the rain percolates slowly down through the fine volcanic ash, where it is filtered, and emerges at Mzima some two decades later as crystal-clear water gushing out at a million gallons an hour. The spring bursts from the base of tumbled black lava blocks, laid over the bright red soil that is Tsavo's signature. There it feeds a series of pools dotted with islands of feathery reeds and fringed with huge yellow-barked fever trees, magnificent newtonia hardwoods, spreading figs and three species of palms. The springs were home to a pod of thirty hippos and a number of crocodiles, and teemed with fish that were hunted by kingfishers, darters and cormorants. Sykes and vervet monkeys, baboons and bushbabies fed on the abundant fruit of fig and water-pear trees, while monitor lizards and snakes hunted among the lava rocks and reed beds. After a mile the pools suddenly disappear under a wall of lava, with the water emerging further on as a slender stream, leaving that mile-long string of pools as a complete green ecosystem in the midst of the red semi-desert. In those days I flew a little Piper Colt two-seater plane and from a height the pools looked like a string of emeralds on a rumpled sheet of dusty red velvet. With my young new assistant, Richard Gichuhi, I made a comfortable camp in deep shade on the bank of a lower pool and started to unload our collection of strange gear.

I had designed what we called the coffin, a coffin-shaped, open-topped metal tank with a plate-glass window at one end and flotation drums on either side. The idea was that Joan would lie in the coffin and I would punt her around while she filmed through the glass. I have no idea, now, why I thought the hippos would accept this gross invasion of their pool. It seems obvious that they would either be frightened and run for it, or be provoked to attack, as they so often do with fishermen's canoes. Standing up in this unwieldy craft and poling it about I was obviously a frightening apparition to the animals, so in order to reduce our profile I walked around in the five-foot-deep pool instead, pushing the coffin ahead of me. A young white boy had been killed by a crocodile here some months before and an Asian man, cooling his feet in the water, had lost a leg after his friends had pulled him from the jaws of a big croc. We were very aware of the danger and, with most of my body underwater and Joan unable to check for crocs except straight ahead, the few trips I made in this fashion were scrotum-shrinking endeavours. The hippos, fortunately, were more frightened than aggressive but the few shots we got of them this way were simply not worth the risks. In the end it was a relief to declare gracefully that the coffin did not work.

My next experiment was a four-foot cube of metal mesh, which I had designed as a crocodile-proof cage. It had gaps in the mesh through which I could film and holes in the bottom so that I could walk it around. This device only just worked on land, and when camouflaged with reeds and in the strong Mzima current it simply got swept over and tipped me out. It was an even bigger failure than the floating coffin. Both devices provided some amusing footage of the presenter type in case we needed it, but were basically useless, and I was worried about how I was going to get the underwater sequences I needed – and had promised.

While I pondered these problems I concentrated on getting some footage from hides I had placed around the springs. Back then only a small number of tourists visited Mzima, and any who did come were restricted to the top pool, where I had a well-camouflaged hide positioned to film animals coming down to drink. One morning a large bus pulled up and disgorged about thirty nuns who were obviously on an outing. They must have been on the road for several hours as almost all of them clearly urgently needed a pee, and as there were no facilities provided they headed en masse for the area hidden from the car park by some thick bush. The very bush that concealed my hide. They obviously felt safer in a crowd, and hoicking up their habits they squatted in sisterly synchrony. Evenly spaced over a large area, those crouched black-and-white figures looked exactly like a colony of incubating penguins. They posed no threat to me but never have I been more frightened that my cover would be blown. I didn't move an inch until every last one of those vestals had departed the colony: only then could I breathe freely again and rejoice in the varied vistas afforded by a well-placed hide.

Before long I had taken plenty of film of animals coming to drink at the pools and a lovely flowing sequence of an emerald-green snake hunting a cream-coloured reed frog, which had leaped for safety, swimming urgently with its orange legs pumping, across the glassy pool. In one of the smaller crocodile-free pools, I also filmed some good underwater scenes of cormorants – silvery bubbles streaming from their feathers as they dived to chase and catch fat, shining fish. Terrapins, their shells coated with algae, paddled by and a big python moved sinuously through the underwater world. A large eel scavenged a dead fish, taking a firm grip, then spinning vigorously until it tore off a morsel. It was all good material, but if this film was going to produce the bang I had in mind I had to get underwater with hippos and crocs. I spent a lot of time carefully watching their behaviour and assessing my options. Crocs and hippos kill more people in Africa than any other species, and going underwater with them was courting danger. Nevertheless, I wanted these pictures, so had to accept the possibility of death, think about how best to evade it, then put the whole issue out of my mind. It had to be done.

I ran my thesis past Joan: 'Hippos recognise man as a vertical creature that they only see above water – on land, on the water's edge or in a canoe. That is their mental picture of man, the image they will attack, or from which they will flee. But if they

are presented with a *horizontal* creature, *under*water, where they probably can't see very clearly, I believe they will simply not recognise it as a man. Hippos have no enemies underwater except for very big crocs, so they may be curious about this strange creature, but will have no reason to attack. I won't use an aqualung because the bubbles could upset the animals, I will just use a snorkel. Ergo, I will be safe if I dive with hippos.'
QED.

'And the crocs?' Joan asked penetratingly. (I'll admit I hadn't got the croc element worked out yet.)

'Well, they are probably the same, but if I slip in quietly when it's hot and they are out of the water sunning themselves the question won't arise.'

'And if it does arise?'

'Er, you can be up on the platform wearing your Polaroid glasses with a clear view of the action. If you see a croc coming you can throw rocks,' I quickly improvised.

'At the croc or you?'

'I guess between us, so the croc is scared off and I am warned of danger. Um, not that it's going to be dangerous as I'll only go in when they are all sunbathing.'

Joan had heard spurious reasoning of this sort before, so I gave her my most winning grin to cover my uncertainty.

'What about when you need to come up to breathe?' she asked, for my appearance on the surface could trigger an attack.

'I will come up only when they have their heads underwater so they won't see me on the surface.'

'What is this film you are making, an aqua-musical? A dozen hippos are going to do synchronised swimming so you can snatch some air while they are all down? You're crazy!'

My winning grin and smouldering allure had obviously had their usual lack of effect, so I just mumbled, 'David A said the films had better be good, so I have to do something big. I'm sure this will work, so I'm going to try it.'

Joan was obviously as worried as, deep down, I was, but I felt sure she had seen a grain of logic in my idea, and she gave me a resigned look I'd seen before and said, 'Well, at least we have a handy and appropriate coffin. Would you like your head or your feet at the window end?'

The next day, after a careful check to make sure the two biggest resident crocs were out on the bank downstream enjoying the late-morning sun, I hoisted the camera in its heavy underwater housing, reminded Joan that 'hippos only eat grass' and lowered myself into the water at the head of the spring. I had dived in the clear seas of the Galapagos and Australia, but nothing had prepared me for the clarity that enabled me to see the group of hippos some fifty yards away in extraordinary detail, the refraction from my mask making them appear much closer. The water was as clear – and as intoxicating – as vodka.

Coral reefs are a riot of vivid colours, with fish in clashing combinations of hues. Mzima was a study in muted tans, browns and olives, and all the fish were monochrome. Shoals of inquisitive, silvery, foot-long barbus swirled around me, while the cylindrical pewter labios, with their large swollen lips, ignored me and kept rasping away at the rocks, leaving a track like a lawnmower's where they had scraped off the algae. The pool is only about five feet deep, so floating just below the surface, I moved closer and closer to the hippo, some lying on their bellies with their heads below the surface, others standing and checking topside. As I moved closer they turned towards me, their eyes open underwater, but since they don't have specialised underwater vision I assumed they were not getting a very clear picture – probably like my own would be underwater without goggles. Their livid pink bellies gave a touch of colour to the scene and on the male in the group several long scars gleamed pale in the clear water. The way they held their heads – with necks arched – showed they were tense and a couple of small calves burrowed nervously into the middle of the massed adult bodies. The current was pushing me towards them and to stop being swept too close I had to lower my legs to stand on the bottom. In my excitement I had quite forgotten that I was meant to be horizontal and sure enough, when they saw me standing they became uneasy, but turned and moved away. Now for the first time I saw their balletic underwater movement as these three-ton animals moved off like men bounding on the moon, weightless and graceful, jumping a sunken log in slow motion, landing with a puff of detritus and disappearing into the reed beds.

I had taken no pictures, but I now knew my thesis was going to work. Shaking with excitement and relief, I swam slowly back to where the spring emerged from the lava. Off to one side of the pool was a scatter of bones – with a large pelvis and curved ribs – all shining white; I saw more species of smaller fish, crayfish and a crab, and the splash as a kingfisher speared a barbus. I had been thinking for weeks about the film's theme, the thread that would tie all the elements together into a coherent story. I had a list of creatures that I knew would feature, but had tried not to have a preconceived idea of the narrative: I wanted it to be shaped by the discoveries we made during the filming. What made these pools so rich in life? Then, as my feet stirred what looked like a mulch of chopped hay lying six inches deep on the floor of the pool, I began to understand. I was stirring up hippo dung. Hippos leave the water at night to graze, each eating up to a hundred pounds of grass, and during the following day they deposit huge quantities of dung into the water. This was the source of the richness and diversity of these pools. Occasionally a hippo would die in the water, injecting a vast amount of extra energy into the system. Death and dung, then insects, fish, crabs, eels, cormorants, kingfishers, terrapins, otters, pythons and crocodiles: the whole pyramid of life in the springs rested on the base of half-digested grass clippings that covered the bottom of the pools. I had found a powerful ecological theme, which would lift the story to a level not seen before on television. I still needed to get back in with the hippos – and the crocs – but with the theme now racing around in my head I was prepared to risk anything.

Over the next few weeks I managed to get closer and closer to the hippos, approaching from downstream so that I wasn't accidentally swept towards them. I found they used regular trails underwater and that by waiting close by I could film their stately, weightless dance as they drifted past. Stationing myself near a tree trunk that lay on the bottom, I filmed them clearing it with a leap as graceful as an impala's. Individuals would come slowly to a certain spot every day, adopt a very stretch-legged stance and open their mouths wide. This was an invitation to shoals of fish to clean around their lips and teeth. In the ocean, I had seen large rock cod come slowly to a station and invite cleaner fish to clear them of parasites – the similarity between these ritualised movements was uncanny. Unfortunately, while having this treatment the hippos were always more sensitive to my presence and I was never able to catch it on film.

But it had to happen, didn't it? One day, as I was filming a hippo with a wonderful shadow cast over it by a papyrus-like reed, a rock splashed into the water just behind me. I spun round and thirty feet away was a croc – about a seven-to-eight footer – his powerful serpentine tail pushing him slowly towards me, his front legs clamped to his sides, a V-shaped line of ripples moving out from his nose. Taking as deep a breath as I could, I sank slowly and sat on the bottom, keeping the heavy camera casing between me and that approaching armada of teeth. The croc came to within about ten feet (which looked like six through my goggles) then, spreading its webbed front feet and facing them forward like air brakes, it came to a stop and, like me, sank slowly to the bottom. We sat there looking at each other. The blood was pounding in my ears, but somehow I felt I understood what was going on behind those unblinking green-gold eyes. Despite having been very successful creatures for perhaps two hundred million years, even a big crocodile has a brain the size of a walnut. It doesn't need more brain power to manage its usual routine, so when presented with something new it is very cautious. I had learned this from their reaction to my well-camouflaged hides here at Mzima, but also on the Nile and other rivers. Presented with a completely unknown creature as big as himself, this fellow was taking it slowly. He had plenty of food in the springs, so he didn't have to take any chances. I guessed that if I waited too long he would become bold enough to investigate. But I couldn't wait, I needed to breathe. With the heavy camera in front of me I made a lunge towards him and he swirled round and was gone in a great cloud of hippo dung.

This was the reaction of all the crocs I met thereafter. If I splashed on the surface as if in trouble, they would come for me at speed, but the moment I sank and sat still they were puzzled, cautious and ultimately frightened. That said, I realised that it would be a very different story with one of the massive twelve or fourteen footers that lived in the lower pools. Eighty or more years old and enormously experienced, a croc that big would have killed many different animals, perhaps some as big as buffalo or young hippo. I would have been an interesting new item on the menu and I doubt he would have hesitated as the younger crocs did. Big crocs spend most of their time resting in the sun, so every day Richard Gichuhi would watch over the two downstream monsters that I feared and radio Joan the moment one entered the water.

Joan started off just watching from the bank, but explaining to her what it was like down there wasn't enough for her: she had dived with sharks and sea lions, and soon she accompanied me into the pools. She was fearless and enjoyed taking stills with a little waterproof Nikonos. One day she was on the edge of the pool reloading film, when a tourist approached and asked her what she was doing. Joan was shy, but thrilled with what we were achieving, so she and the old gentleman had a long talk about our project. He asked her if the Nikonos was any good. She said she would rather be using our big Hasselblad camera, but that they didn't make an underwater housing for it. 'We are making one now,' said her admirer, 'and I would very much like to send you one. I am Victor Hasselblad!'

On another occasion, when coming up to breathe, I saw Joan laughing with a tall, tanned, good-looking, bare-chested fellow, all washboard abs and rippling pecs. I speedily scrambled out of the pool to check on what was going on. It turned out to be Wolfgang Weber, a young German wildlife artist who, tired of waiting for me to reply to his letter asking if he could join us, had decided to come to Africa anyway. He accompanied me many times underwater, making quick sketches of hippos and crocs with waterproof paper and pencils. Wolfgang became a great friend and his wonderful drawings grace this book.

One day there was a great commotion in the top pool and we saw that a croc had caught an impala ram and pulled it into the water. I obviously couldn't get straight into the water, but we still had the coffin, and we poled it into position so I could film through the glass as several crocs shared the kill. Their peg-like teeth are unable to shear through meat or bone, so they feed by taking a grip on a piece – a leg or even the head – and spinning over and over, pulling against each other to tear it off. Seen through the glass from just a couple of feet away, the power of those spinning bodies was formidable and it made for a spectacular and primeval sequence, with flashing teeth, flying spray, bubbles and blood in the water. Just as feeding lions are joined by vultures, here shoals of fish swooped in to snap up scraps, and several wary terrapins edged in and tore at the carcass like jackals.

By the end of our time at Mzima Springs we had captured all the images I had hoped for and had a great story to tie it all together – the pyramid of life built on hippo dung. I had started with a bang as planned, but it had almost wiped us out. Fortunately, *National Geographic* took an article using lots of our pictures, with Joan as their cover girl, underwater with a twelve-foot python. The film also raised enough funds to pay for a large, circular underwater observation tank, with windows all round, which is at Mzima to this day, allowing visitors a glimpse of the world that we had got to know so well.

Ten years later we revisited Mzima to shoot how we had filmed the hippos the first time. We dived with aqualungs and a great cameraman, Martin Bell, was in the water with us. We gathered much better footage than we had achieved with my old underwater cameras and even some footage of Joan scratching a hippo's bottom in imitation of the cleaner fish. Then, one day, just upstream from us, two males got into a fight and stirred up so much detritus that we lost all visibility. The safest thing to do was to sit on the bottom and wait for the water to clear. The three of us were nervous, breathing hard and producing lots of bubbles. As I couldn't see anything, I didn't realise that one of the males had come very close to us. Seeing the bubbles from our aqualungs, he must have thought it was the other male coming for him again. He could see as little as we could through the turbid water, but he charged with open mouth, scything through the murk with his eighteen-inch canines. Joan was knocked flying by the pressure wave ahead of its great body and lost her face mask. Then the hippo bumped into me, got my right leg in his mouth and shook me like a dog with a slipper. The people on the bank said I was waved in the air three times and slammed on the bottom of the pool so hard that my back was bruised and my swimming trunks full of gravel and hippo dung.

Anyone who has been mauled by a big animal will tell you that there is no pain at the time, just a numb awareness of what is happening. I knew that my calf had been impaled by his left canine, and my foot was over to his right, being crushed by the rear molars on that side. Hanging down from his mouth, I could feel the hard edge of his lower lip and his chin whiskers rubbing the back of my thigh. He shook me up and down vigorously, then let me go and stormed off. I had lost my goggles, mouthpiece and camera, my air-tank straps were torn off, and I just stood there for some seconds in shock. I suddenly realised that the commotion would attract the crocs, and I was pouring blood. Joan was only about five yards away, but in the muddied water had not seen what had happened and was rather dazedly looking for her mask. 'Get out!' I yelled at her. 'Go, go, get out – NOW!' She headed for the bank and then something grabbed my arm with a painfully fierce grip. 'No!' I shouted, and spun round to find Martin. He had been sitting close by, unable to see but aware of the tumult. When the water cleared all he could see was blood, and he came to the rescue and pulled me to the shore with the power of a salvage tug.

There was a huge hole right through my calf and now it started to hurt. A tourist who had seen the whole thing, an excitable Italian doctor, bound my leg tightly with a crêpe bandage: so tightly that he cut off the circulation and when we finally made it to the doctor in Nairobi – courtesy of a tourist's plane from the nearby lodge – he said that if we hadn't got there so quickly I would have lost the leg. When the crew went out later with a boat to collect my camera, they found Joan's face mask. The glass was broken and there was a two-inch gash through the rubber. One of those huge slashing teeth had come within an inch of her face and she had been completely unaware of it, thinking the mask had just been swept off by the force of the hippo's bow wave. When she visited me in hospital and slipped it on to show me I went very white. Another few inches and her head would have been split in two: the words pumpkin and pickaxe came to mind!

I was lucky to be in the care of Imre Loeffler, the brilliant Austrian surgeon who had amputated my finger ten years earlier. Imre had seen many hippo bites during his years in a bush hospital in Uganda and had a very basic technique for treating animal wounds, which are always messy. My fibula, the smaller of the two bones in the lower leg, was broken, and the wound was full of detritus and the chopped grass that is hippo dung. In his thick accent Imre told me, 'I vont you to vigorously vosh the vound.'

So every day I would slosh the leg back and forth in a warm bath to wash out the foreign matter and necrotic tissue. The bathwater would soon turn pink with blood and I would lie back with a drink, reading the current bestseller, *Jaws*, and entertaining selected visitors by pushing a coke bottle through the hole in my leg.

Our first solo production, *Mzima: Portrait of a Spring*, was a great success, so it was with some confidence that we moved our camp a few miles from the spring, to a hidden valley filled with the subjects of our next film, baobab trees. The baobab is Africa's grandest tree and the world's largest succulent plant. Grotesquely squat and massive, its enormous branches look like a spreading root system and some African legends say that

God, having a bad day, pulled up the baobab and planted it upside down. Like elephants, they are the right scale for the vastness of Africa – wrinkled silvery-grey giants, fifty to sixty feet in diameter and not much more in height, they rear above the dry and leafless bush. In Tsavo, the bottom ten feet of their trunks are stained red, where generations of muddy elephants have scratched their itches. Higher up there is often a fringe of short, up-curled strips of bark – like the paper ruffle round a fancy lamb chop. These are the broken ends of long strips of bark torn off by elephants and they can still be seen, like stubby accusing fingers, on trees in areas where elephants disappeared a hundred years ago.

Baobab trunks are often hollow, and can collect hundreds of gallons of rainwater. With a hole drilled in the trunk and sealed with a wooden plug, the tree can provide water for several months – such trees were considered so important that hundreds of them were marked as 'water points' on early British maps of the Sudan. Other hollow trees have been converted by man into guest rooms, pubs, prison cells and bus shelters, but we were interested instead in all the uses wild things had for the tree. As we were going to be camped beside one for perhaps a year, we spent a week searching for the best campsite, one with shade and a good view, and close to our subject. If there was water nearby that would be a bonus. Before long we were under canvas on the morningside of a particularly fine specimen that we had chosen as the star of our film and we started to observe the baobabs. The great documentary maker Robert Flaherty once said, 'All art is a kind of exploring. To discover and reveal is the way every artist sets about his business.' This is particularly true of wildlife filming, when every new project means learning to see all over again, to pick up the clues and signs that lead to an understanding of this new and different territory. Today it means googling the subject and getting buried under a mass of information; back then it meant looking, listening and finding out for yourself.

Walking through the bush together, Joan and I found that our eyes went to different parts of the trees. Mine penetrated the foliage, spotting nests, holes and snakes inside the cover; Joan's curiosity was sparked by the outer vegetation, noticing insects, chameleons sunning themselves and flowers. There was one sequence I had set my heart on: several species of hornbill breed in baobabs and they have a unique nesting procedure. Most species are black-and-white, long-tailed, pigeon-sized birds with long, deeply curved beaks. The female squeezes through a hole into a hollow branch, then her mate, making many trips to the nearest source of mud, plasters up the hole, leaving only a vertical slit half an inch wide and some two inches long. Inside, safe from predators, the female will incubate her eggs and care for her young, while the male makes thousands of trips to bring her insects, small lizards and other tasty items. This much was already known about hornbill behaviour; in fact, my old mentor Myles North had written it all up decades earlier, but I wanted to film what had never been seen – what went on inside the hollow tree – and to do that I would have to put a window into a nest. It would be too disruptive to do so once the female was blocked up inside, so we searched for holes where hornbills had nested before, and which they were likely to use again. We recognised them from the traces of plastering around the holes. We found two nests that were close enough to the ground not to require us to use a tower to film, so I prepared them both for future photography. With a keyhole saw I cut out a rectangle from the side of the nest hole and fixed it so that it could be unscrewed easily and removed. A piece of plate glass was cut to fit the window, then the slab replaced and the nest left undisturbed. Now we'd have to wait until the next rains, when the birds bred, to see if any hornbills would move into their old, but tastefully renovated, quarters.

Like islands in this scrubby bush country, baobabs attract many species of birds that nest in their branches and hollows. Several of Africa's most beautiful birds are found here, and the species we filmed nesting in the tree were the stuff of bird-watchers' dreams: orange-bellied parrots, lilac-breasted and rufous-crowned rollers, red and yellow barbets, striped king-fishers, green wood hoopoes, golden-breasted starlings, red-beaked and grey hornbills – some of them involved in noisy and colourful territorial battles over a particularly good hole. Inside the hollow trunks little parties of bats hung next to the nests of spine-tailed swifts – tiny cups of feathers and saliva – while spotted eagle owls peered out through the larger holes. On top of our chosen tree a Wahlburg's eagle had built its substantial stick nest and in the branches were collections of twigs carried there by red-billed buffalo weavers, who barricade the approaches to their nests with thorns, to protect them from intruders – all except snakes.

Our film was becoming very bird-heavy and we still had the long hornbill sequence to capture. I needed more variety and we found it in some spectacular insects. One was a beautifully camouflaged, ten-inch stick insect, which was so realistic that we filmed a buffalo weaver try to pick it up to add to its stick barricade. We found a large praying mantis called the Diabolical Idol eating a baby gecko, holding it firmly clamped in her long, grappling-iron front legs. When disturbed by a bird landing nearby, she whipped her legs straight up on either side of her head, turning them flat-side forward, to flash a large blue eye with a red circle in the middle around that triangular, alien head. The shock effect was dramatic and the bird took off with a frightened squawk.

Anticipating the rains, the tree was soon hung with large flowers on long stalks, white, waxy petals and a rather sickly-sweet aroma that fills the air around a big tree as they open just after dark. When I climbed down from my filming tower at the end of a long day I often brought with me a couple of swollen buds, which Joan put in a jar on the table to burst open as we had dinner. Despite my primitive equipment I was able to film this action in time lapse. The flowers are pollinated by epauletted fruitbats, with foxy faces and big intelligent eyes, and several of them visiting the flowers had young clinging to their upside-down chests. Tiny Senegal bushbabies also helped pollinate the flowers – hamster-sized, huge-eyed, long-tailed creatures that periodically pee on their hands and feet in order to scent their passage, they make prodigious leaps from flower to flower, their little elfin faces yellowed with pollen. The misshapen moonlit branches were alive with strange cries, rustlings and flutterings, and it was easy to see why African legends say that the baobab is haunted by spirits. The flowers are short-lived, turning brown and falling to the ground within twenty-four hours, to be slowly replaced with large green fruit capsules that dangle from the branches like Christmas tree decorations and are full of powdery white pulp that, stirred into water, makes a tasty drink. A troop of baboons visited all the trees in our area to feed on this appropriately named monkey bread.

One morning I was in a hide near the nest of the Wahlburg's eagle when Joan called up to say that the red-beaked hornbills had started to move in. We filmed the male flying in with load after load of mud, sealing his mate inside and bringing her insects, but for now that was all I was prepared to do. The female's attachment to the nest at this stage would be weak and disturbing her would risk her deserting the eggs. I would have to wait until she was well into incubation before doing any more filming. A chicken's egg takes three weeks to hatch; I would wait about fifteen days before I dared open the hornbill's hatch. In preparation for that day I erected a canvas hide thirty yards away, and as the male became accustomed to it I moved it closer and closer until it was up against the side of the tree and fixed to the bark around the hatch. The hide was lined with black cloth so that when I did replace the hatch with glass the female would not be able to see me. A single photographic light could then be brought up slowly on a dimmer switch to enable me to film. Fortunately, hornbills prefer a nest hole that extends up within the tree above the nest cavity. When the female is disturbed she climbs up into this hole to hide. So when the time came, I tapped on the hatch until I heard her climb up into her refuge, then quickly replaced the bark cut-out with glass and waited.

It was a very tense few moments. If she came down and panicked at the glass she might break the eggs in her scramble to escape, or she might climb up into her sanctuary and just stay there. I had to trust that I had taken it *polé-polé* enough and that she would accept the changes. The male was very used to us by now and Joan had thrown him a big grasshopper that he brought to the hole – which we hoped would reassure the female that everything was completely normal. Her reappearance was dramatic, for the hollow was slippery and she simply plopped down on to the eggs with a squawk. She accepted the grasshopper, pecked curiously at the glass a couple of times, then maternal instinct kicked in, her brood patch feathers fluffed out and she slowly lowered herself on to her eggs, shuffled a few times to get comfortable and closed her eyes. I wanted to yell in triumph but had to whisper into the radio to Joan, 'The hornbill has landed.'

The next six weeks in that hide brought a series of surprises, discoveries and sequences that provided a unique and ongoing thread through the film. The first surprise was that the female was half-naked and sat there surrounded by fallen feathers, her wings like flippers and her tail a comical two-inch stump. Being cloistered in the nest gave her an opportunity to moult all her major feathers at once, while in the world outside her mate would be replacing his flight and tail feathers one or two at a time so as not to compromise his flying abilities. Another surprise was that the female collected the chick's faecal pellets and used them to plaster the hole from the inside. Unsurprisingly, she treated the window as another, bigger hole and constantly smeared the glass with faeces. I had quickly to cut more sheets of glass, and Joan or Gichuhi were kept busy cleaning them. A tap on the glass would send the female up the hollow so I could quickly replace the window, but I did miss several shots because of the filthy glass. She had laid her eggs two days apart and, because she had to incubate from day one, they hatched at that interval, so with three eggs the oldest chick had a considerable lead on the youngest. Being bigger and stronger, it grabbed most of the insects that the male pushed through the hole, so the size gap between the chicks increased. The female had grown a full suit of new feathers and with the chicks growing rapidly the nest was getting really crowded. To save space all of them sat very upright, keeping their long tails pointing straight up their backs. Closed inside my cramped hide, crowded with my equipment and looking out through a small hole, I felt a bit like the jam-packed birds on the other side of that window.

To get shots of the male pushing through the slit I needed warning of his approach so the camera would be running when he arrived. Every day Joan sat under a nearby tree taking still pictures, notes on the food brought and providing invaluable information on a walkie-talkie. 'Male coming in with a mantis', 'Ooh, you must get this, on his way with a big gecko', 'Ten minutes before it starts raining', 'Coffee and bikkies ready when you are'. Three weeks after the eggs hatched the female was ready to break out and she began hammering at the mud with her beak. Frightened by this new activity, the chicks huddled at the back of the nest. She chipped away at the hole all day, accompanied by great excitement from the male. After feeding four mouths for so long, he needed the help she would provide.

When the hole was large enough she squeezed out and, after six weeks of being unable to stretch her wings, dropped straight into flight. Then came another surprise: as soon as she had wriggled out, all three chicks went to work to reseal the hole. With rapid sideways drumming of their beaks, using their own droppings and mud brought by the male, they had closed it back to a narrow slit within a few hours. This was all new information and we spent many an evening round the fire wondering why the hornbills had evolved such a complex system. Nesting in the very same tree were at least six other species who raised their family in a simple hole, apparently no less successfully than the barricaded hornbills.

After another three weeks the chicks had grown their full plumage and one day the biggest started to chip away at the hole. This triggered a fascinating dichotomy in the behaviour of the three chicks, for the two younger birds were not ready yet – they still had several days to go before their inner clocks told them it was time to break out. Cowering in the back of the nest, they watched their sibling's strange behaviour with what looked like horror. Every time the oldest chick took a break from his tiring job the other two youngsters quickly moved forward, picked up the chippings and stuck them back, using their droppings as glue. With one bird breaking and two rebuilding this behavioural battle went on for two days. Then the middle chick's alarm went off and it switched from repairing to chipping, enabling the oldest finally to break out and make its maiden flight. The middle chick followed, and after two days of solitary repair work the smallest changed tactics, broke down his plastering and wriggled out to join them. We had caught on film an amazing piece of natural history.

Finding a great shot with which to bring a film to an end is always a challenge and I was still not sure what this might be. The problem was solved for us by the prolonged dry season that hit Tsavo that year. In the search for moisture, elephants were using their tusks to dig into the baobabs and chewing the juicy, pulpy wood, sometimes digging a hole right through the huge trunk. Many trees were so weakened that they came crashing down, and we found one that had fallen and crushed the elephant that had been tusking it, the two giants of the bush country dying together in a monumental scene that gave us a dramatic ending for the film. It got some wonderful press, the best line being, 'Only God can make a tree, the Roots made *The Baobab.*' Is that equal billing?

8

Lassie Come Home

After I had completed the film about our bongo capture operation – the third film in our BBC contract, instead of Lake Naivasha as we'd planned – David Attenborough asked me to stay on. Armand Denis had now retired and his popular *On Safari* series had come to an end. The BBC needed a new wildlife programme presenter to take his place – his mantle awaited me. I didn't think I would look good in a mantle, I had proved that a presenter was not necessary and neither Joan nor I had ever been interested in money for its own sake. What we wanted was the ability to spend more time on our projects and to afford the equipment to realise them. I wanted to make the definitive film of the Serengeti migration, which would take at least two years, and involve aerial photography and thousands of miles of expensive, cross-country, Land Rover-based work. I also wanted to spend another two years filming the life within and around a giant termite mound – another complex ecological story that would require specialist equipment. So yes, we did need more money and I headed across the Atlantic in search of it, with *Mzima* and *Baobab* in two large cans under my arm. I was off to crack the American market.

Foreign-made documentaries simply did not get on to the big three American networks that dominated US television at the time, but fortunately, because a hick from Africa sitting in the waiting room clutching a couple of tins has a certain novelty value, I was welcomed into the offices of the top documentary honchos of the time. Ivan Torrs, who produced the children's TV series *Flipper*, wanted our Mzima Springs film, but he also wanted to edit in sharks and alligators to augment the 'jeopardy' factor, which I soon discovered was a much sought-after commodity in these circles. David Wolper, probably the most successful documentary producer of all time, took my films home, viewed them in his own cinema and the next day gave me another insight into the US market. 'I loved them! I want to buy them both – they are different and break new ground. But you've gotta realise that this is commercial TV, many of the advertisements are for food products like Kraft cheese or Ritz crackers, so we're gonna have to get rid of a lot of the shit.' I didn't think any of my material deserved that description, but he went on, 'You've got a crystal-clear spring – beautiful – but it's full of hippo shit. Kraft aren't going to like that. Then you've got the hornbill smearing mud 'n shit all over the place to seal its nest. They're not gonna like that either!' Despite his concerns he offered me eye-watering prices for the rights to both films. I'm no businessman, but I knew I was completely out of my depth and liable to be taken for a ride, so I said thanks, I would get back to him and took my cans home to Kenya to talk it all over with Joan.

The money would enable us to do what we wanted, but I was upset that the films had been seen as commodities that could be tweaked to suit the market. Take 'the shit' out of the Mzima film and the whole story falls apart. I had always been aware that our films had to have an audience, whom I strove to amaze and entertain, but apparently now it would be necessary to please the sponsors first. I couldn't imagine having to make a film with the needs of Kraft cheese in mind. Then a piece of good luck came our way. Before we had made a decision, a telegram arrived from Colin Willock, the director whose decision to exclude me from the editing of our Galapagos film had led me to leave *Survival*. It said simply: 'CONGRATULATIONS ON *MZIMA* AND *BAOBAB*! LASSIE COME HOME!'

Within days Colin had flown out to Kenya to explain that *The Enchanted Isles*, our Galapagos film, *had* just cracked the American market – the first British-made one-hour wildlife special ever to do so – and that they now wanted more, ideally from the team that had made the Galapagos film. *Survival* would take our two films, Colin said, we would be fairly paid and they would use their clout to make sure the films were not sanitised for the sponsors. Not only that, but we would get the budgets we needed for the new films I wanted to make and would also have complete control over their production. As Colin returned to London with the cans under his arm, Joan and I geared up for the next incredible decade. We were going back to the Serengeti.

The Serengeti is my favourite place on earth, so it was wonderful to be back among my friends there and the great thundering herds on the plains. This would be a life-cycle story – following the yearly migration of that unlikely-looking antelope, the wildebeest. Famously described as being designed by a committee and assembled from spare parts, these dark-grey, horse-sized, slender-legged creatures are, in fact, incredibly successful animals that once occurred in immense migratory herds in many parts of east and south Africa, but now exist in huge numbers only on the Serengeti. We arrived in January, just in time for the birth of the calves, which takes place out on the eastern short-grass plains.

These vast flat plains were created by the volcanoes of the crater highlands to the south-east that erupted in the Pleistocene, covering the land downwind with layer upon layer of ash. One layer calcified into a hardpan and was then buried under more ash, so that it now lies about a metre below the surface. Except where erosion has cut down through the crust, trees do not grow here, their roots unable to penetrate this concrete-like barrier. With the rainfall they receive these plains should evolve into woodlands but, locked into stasis by this hardpan, the climate, vegetation and animal life have remained essentially unchanged for millennia. The short grasses that grow here are rich in potassium and calcium, providing a perfect diet for the lactating wildebeest females and also, importantly, no cover for predators. Females about to give birth form groups, usually at dawn, giving the newborn calves a full day to gain strength before night falls and the danger from predators increases. They often lie down and start calving simultaneously, the pale brown calves slipping out on to the grass after a few minutes of labour. The cows lick the baby clear of the foetal membranes, while the calves struggle to rise, shaking their heads vigorously to clear the fluid from their ears. In another ten minutes they struggle to their feet, stumbling, staggering and falling time and again until they steady themselves and take their first drink. Ten minutes more and they can run. By the end of the day they will be as fast as their mother. By synchronising the calving in this way, producing half a million young over two to three weeks, the wildebeest flood the food market. Predator numbers are regulated by the food available throughout the year, and for most of the year the plains only host small numbers of gazelle and warthogs, so there are few resident predators. At calving time, more predators move out on to the plains but, almost overnight, there are suddenly so many newborns available that the carnivores are soon sated. Many, including jackals and striped hyenas, can live off the wilde-beests' nutritious afterbirth, which is plentiful on the plains, and for a while do not need to hunt at all. After a couple of weeks the calves are no longer such easy targets and the bonanza comes to an end.

By May the plains have begun to dry out. Rain begins to fall to the north and west, and now the herds coalesce and head in that direction. Do they move because the available water on the plains has become too saline? Are they heading towards the smell of rain, or the lightning they see at night? We still do not know and I prefer it that way. I find it deeply satisfying that we do not understand why over a million animals suddenly move hundreds of miles – there are things in nature that are best left a mystery. May or June is when the herds are at their most spectacular, for when they reach the woodlands at the north-western edge of the plains the front row of the vast gathering slows down. They may pile up here for days, fearful of the bush and reed beds, and the predators that could be lurking there. There are no leaders, just a slow, random forward movement, and it is often one or more zebras, who travel with the wildebeest, that break the logjam and then the herds pour through the gap in the reeds the zebras have shown to be safe. But of course it is not safe. The woodland prides of lions have been waiting for the arrival of this annual bounty and the older lionesses know just where to lie in ambush. Streaking out from the cover that the wildebeests rightly feared, they plunge like golden arrows into the choking clouds of dust and milling hooves. The herds scatter and gradually the dust settles to reveal a lioness with her kill. She stands astride it, panting heavily, then makes a low moaning call and out of the reeds come her little cubs, lolloping across the pulverised earth.

These are the enduring scenes that every year mark the passage of the herds from the open plains to the woodlands and beyond. They move north-west now, down the long corridor that extends almost to the shores of Lake Victoria. Then the rut begins: a frenzied couple of weeks during which a quarter of a million bulls compete to mate with more than three times as many cows. The bulls claim and defend small territories where, racing back and forth in clouds of dust, they try to keep some females long enough to service them. But the herds are on the move, so the bulls, too, are in ceaseless motion, patrolling their patch, fighting off rival males and trying to stop the females from moving on – grunting like giant bullfrogs, rearing up on their hind legs and pirouetting, dropping to their knees to clash horns with the neighbouring bull and stopping only for a brief coupling with any cow who will stand still for a moment. The bulls live up splendidly to their nickname: the clown of the plains.

By July, far to the west, the herds are arriving at the shrinking pools of the Grumeti river. Joan and I have been following them, spending much of the day in the Land Rover and sleeping on the roof under the stars. Every day we break for lunch, when the light is too harsh for filming, and take long walks to exercise our legs, looking for future campsites or checking the stream beds for crossing places or fossils. Then we make camp under a big tamarind tree on the banks of the Grumeti – we know what is coming.

In the two and a half years we spent on this film we were the first to document the annual mayhem that takes place in just a few pools along this seemingly insignificant river. Every year the wildebeest come down to these pools to drink, the very same pools that house some of the largest crocodiles in Africa. When we filmed there, in 1972, poachers had been active along the river for years and the Grumeti crocodiles were extremely shy, but we were able to film the tense moment as a huge croc moved slowly towards the drinking herds. As soon as the first animals begin to drink, those behind gain confidence and press forward, often pushing the front row into the shallows. The herds are nervous, but most of them have never seen a crocodile before, so the greenish log moving imperceptibly towards them is usually of little concern. The air is filled with mooings and splashings as the animals push deeper and the 'rat-tat-tat' of a woodpecker adds tension to the soundtrack. The front row is belly-deep in the water now. The log drifts closer to the line of lowered heads, their broad black nose pads pulsing rhythmically as they drink and their white-rimmed eyes rolled upwards to look out over the pool. One animal suddenly jerks up its head, water pouring from its long white beard as it focuses on the log, now only a foot away. It stares for an uncomprehending moment, but its neighbours are still drinking, so it relaxes and lowers its head. Sitting in a hide just a few metres away, absolutely at one with the whole scene, trying to stop trembling, the tension I felt at these moments was intense and palpable. The camera is running, running, running, Oh please don't run out before . . . and suddenly there is an explosion of spray as an unexpectedly huge, scaly olive body arcs upwards in a flash of gaping jaws and teeth, a maelstrom of hooves, horns and dust as the herds turn and race up the bank. As the waves subside, a wildebeest steams out to deeper water, its tail waving weakly, its head held under by rows of white teeth.

No wonder some Africans call the crocodile 'the animal that kills while smiling'. Wide V-shaped patterns spread across the pool as more crocs speed towards the kill and soon several have locked their jaws on to a part of the animal. What follows looks like fierce competition for the food, but is in fact a cooperative effort that enables the crocs to feed more efficiently. Each takes a vice-like grip and slowly spins its body two or three revolutions, then gives a violent jerk of its head to tear the piece off. It was just like the scene we had filmed at Mzima Springs, but this time with monsters of fifteen feet and more. With several crocs, some weighing over a thousand pounds, working together in this way, a wildebeest comes apart like tissue paper. Even as the spinning bodies and jerking heads roil the pool and send bloody spray flying, a row of wildebeest heads appear over the top of the bank and the first animals make their tentative way down to the water's edge. The slaughter goes on for as long as the herds come down to drink. The actual numbers killed are insignificant, but it is the repetition, the daily – sometimes hourly – price that the herds must pay to get a drink that is so dramatic.

The last big sequence I wanted for the film was a river crossing. In the late fifties the Grzimeks believed the wildebeest numbered roughly one hundred thousand and since then that number had doubled and redoubled. When we were filming in the seventies, the herd was approaching a million for the first time that century. Their numbers had been climbing over decades after a holocaust hit Africa around 1889 – a deadly cattle disease called rinderpest that was believed to have arrived in Africa with stock brought in with the Italian invasion of Ethiopia. Within three years it had wiped out ninety-five per cent of cattle in East Africa, bringing decades of devastating famine to the Maasai and other tribes, and decimating herds of buffalo and wildebeest.

Even a herd of almost a million wildebeest, accompanied by roughly a quarter of a million zebra, had enough grazing in the north of the Serengeti not to need to cross the border into Kenya, where they would have met the wide Mara river and given me a spectacular crossing. But there were very heavy storms that year and the relatively small Grumeti river was so swollen that the mass crossings made for dramatic sequences.

Flinging themselves mindlessly into the flood, the animals frequently faced a sheer wall on the other bank and would be swept downstream, desperately searching for a way out. When they did find a sloping bank their dripping bodies and churning hooves often turned it into deep mud, in which smaller animals got stuck. Others tangled their horns in fallen trees or hanging vines, rolling their eyes and struggling until they sank from exhaustion. Filming it all on foot, running along the riverbank to get the best angles and seeing for the first time the animals' desperate struggles, was an exciting but saddening exercise. Today, with the herd grown to some one and a half million, the wildebeest have to seek grazing in Kenya, where their epic crossings of the Mara river are considered to be one of the wonders of the world and are watched by large herds of tourist cars.

The birth of the calves, the move off the plains, the many hazards of the trek, facing predators of all kinds, trial by fire or water, and the final return to the calving grounds on the plains were the building blocks of our film. The mortar was provided by weaving in the multitude of smaller species that share the Serengeti with the great herds and whose lives impinge upon the herds or vice versa: the botfly that lays its eggs in the wildebeest's nostrils, from where the maggot will climb up to the animal's brain causing it to walk or run in tight manic circles until it dies or its strange behaviour attracts a predator; the dung beetles that every day roll hundreds of tons of dung into balls, lay their eggs on them and trundle them off for burial, where they nourish both the beetles' larvae and the soil; the vultures that scavenge the thousands of carcasses, and defecate off the high cliffs where they roost and breed, transferring large amounts of energy into the arid valley below; the moth, related to the familiar clothes moth, whose caterpillars eat not cashmere and socks but the horns and hooves of dead animals on the plains.

From my new balloon I filmed the vultures from above as they circled the herds and we concealed a radio-controlled camera inside a big tortoiseshell, which produced spectacular shots of wildebeest and zebra leaping over it. One totally unexpected event provided some of the film's most moving footage. We were filming a dense herd of many thousands of wildebeest with young calves moving towards Lake Lagaja, a shallow soda lake at the top end of the Olduvai gorge – where the Leakey family discovered spectacular fossil remains of early man. As the herd reached the lake shore, we saw a lone male lion come out of the long grass and stand looking at the passing herd. He was too far away to be a threat to them, but his indolent menace caused the lead animals to take fright, and they turned and started to cross the lake.

Most years Lake Lagaja is only three feet deep and just half a mile wide, so the herds often walk across. But heavy rains had raised the level a couple of feet, so they would now have to swim and the calves – only two weeks old – could not swim as fast as the adults. The mothers weren't able to look round while swimming, so when they reached the shore most of them found they had left their calves behind. They turned to face the water, stood in the wind and mooed mournfully; then many started back across the lake towards their young. The caustic water may have made the calves hard to recognise by smell and perhaps all the splashing masked the individual's usually recognised calls – but whatever the reason, the pandemonium just increased, with calves and adults crossing and recrossing the lake in a desperate search, with many passing each other out in the middle. More and more animals poured in off the plains and, being wildebeest, they followed the herds and soon there were six or eight columns swimming in opposite directions.

It was, Joan pointed out, the worst-organised migration we'd ever seen. I waded out into the lake and, by keeping just my head, hands and camera above water, I was able to film the dog-paddling, rolling-eyed animals as they passed in mindless procession. It continued for three days, and when the herds finally moved on the bodies of almost a thousand calves were washed up on the lake shore, and another thousand or more ran through the woodlands in frantic, bleating, motherless herds. This profligacy affected the lives of many other creatures, great and small. A sated male lion, his hugely full belly swinging uncomfortably, plodded towards the shade without ever looking back at the group of lost calves that followed him closely – all desperate for the company of any larger animal. A well-fed leopard slept in a tree, which he had festooned with the carcasses of five calves; marabou storks and vultures tugged at the bodies in the shallows and a week later the lake shore was littered with bleached bones, providing the little brown-and-white Magadi plover with perfect camouflage while sitting on her eggs.

Exuberant fecundity and commensurate waste are not unusual in nature, but are more often observed at the bottom end of the scale. Tiny chironomid flies hatch in billions from African lakes in dense clouds that look like oil fires, locusts darken African skies from horizon to horizon, corals, crabs and barnacles turn miles of ocean milky with their eggs or larvae, and billions upon billions die. The Serengeti is the last place on earth where we can witness this careless turnover of so much life among the larger animals.

As we crossed the Serengeti we made a stop at the Olduvai Gorge, where I had wanted to film the herds moving past a fossilised wildebeest skull. Mary Leakey was in residence and, as always, asked us to stay. She was a wonderful lady – a rapier-sharp mind under a rough-and-ready exterior and with a great sense of humour – totally in love with the barren and baking gorge, and completely dedicated to the work she was doing there. Colonies of swifts had built their nests under the thatch above her dining table and in a minimal sop to hygiene she had hung a large sheet of sacking high above the table to catch the droppings. I don't know how many years that hammock had been there but I do know that it bulged ominously under a huge weight of guano that grew daily. Straws and camels' backs always came to mind, and eating a meal below that sagging sack was an exercise in optimism. After dinner it was always a relief to relocate to camp chairs under the stars when Mary would get stuck into her whisky and cigars, and good conversation would follow.

On one such visit she asked us to accompany her to Laetoli, where she had uncovered a string of fossilised hominid footprints. Her discovery was seminal: it pushed the date for bipedalism back to over three million years ago. Two of these distant relatives had walked side by side on a fresh ash-fall from a nearby volcano and left a set of prints that were as close and evocative as lovers' footprints on a beach. Mary was puzzled by one of the sets of tracks, which had a double heel mark in every print. Several archaeologists and other specialists had examined them, concluding that the creature must have

had a problem walking, which had caused its feet to jump forward a fraction on each step. This just didn't sound very plausible and when Mary showed me the prints I was pretty sure I had the answer. I had seen young chimpanzees and gorillas playfully grasp a bigger animal by the hips and walk behind in follow-my-leader fashion. Human children do it too, often deliberately stepping into the other prints. To avoid bumping feet they must walk in step so the tracks are aligned and quite likely to be superimposed, and the smaller feet would provide that clear extra heel mark. I later sent Mary pictures of children, chimps and gorillas walking together like this, but she didn't need evidence. Looking at those tracks this possibility was so much more convincing than imagining that one of the animals had some strange impediment in both legs.

At Olduvai I filmed the wildebeest pouring through the gorge, pulverising the fossil-rich volcanic ash and, as the film commentary said, 'scattering their history as carelessly as they scatter their dead along the march'. It was exactly the final scene I had wanted. *The Year of the Wildebeest* had taken us well over two years of wandering the Serengeti and following the herds, thousands of hours living with, watching and recording one of the greatest natural shows on earth. Now I was ready to move on to a harder subject, this time with a cast of tens of millions.

Back in the days when I had accompanied my father on safari to visit his cattle buyers, and we had stopped for a picnic and I had had my first taste of beer, I had spent some time examining a termite mound – my dad had even taken a picture of me next to it. I clearly recall that day as the beginning of my fascination with termites and with all the other creatures that make use of these castles of clay.

Termites are often called white ants, which they certainly resemble, but they are in fact more closely related to cockroaches. They exist in unimaginable numbers, eating dead grass and wood, breaking down the cellulose and accelerating the return of nutrients to the earth. In the process of constructing their mounds they bring up minerals from deep in the earth that are then washed downslope to create a large tear-shaped patch so their mark on the land can easily be seen from the air. Agricultural tribes know that these patches make superior gardens and the gold-diggers of Africa's ancient kingdoms also knew that the mounds gave a picture of the precious metals that lay deep below. This knowledge has been picked up on by big multinational mining companies, which now routinely sample termite mounds as the cheapest way of prospecting over large areas. Huge deposits of copper in Mozambique and the world's richest diamond mine, Jwaneng in Botswana, were located with the aid of the lowly termite. The mounds they build vary enormously. In areas prone to flooding they may be great rounded heaps, forty feet in diameter and eight feet high, weighing many tons and covered in large trees, which form islands when the land floods. In semi-desert country they are slender fifteen-foot chimneys, designed to draw out the hot air and all built grain by grain, stuck together with saliva, by tiny creatures that cannot see what they are building.

Like many creatures that live in the dark, termites are soft, white and blind, and in their mindless millions they build structures that are architectural marvels. Smooth tunnels sweeping through graceful arches connect the various chambers, and the mounds have their own water wells and complex air-conditioning systems. Deep inside are nurseries and convoluted gardens made of chewed wood on which the termites cultivate fungi that they eat to enable them to break down the cellulose in wood. Termite society works on a caste system, with classes of workers and large-pincered soldiers, and deep in the heart of the mound is the royal chamber, a cell built like a hardened command post, where the king and queen live long, pampered lives. The king looks no more royal than a worker, only magnified three times larger, but his queen makes them one of nature's most startlingly odd couples. Four inches long, and fatter than a man's thumb, a constant flow of ripples runs down the gleaming ivory sides of her grotesquely swollen belly. Unable to move, apart from these perpetual pulsations, she lies there like a half-inflated airship as dozens of tiny workers stroke and clean her bloated body, many hanging from the ceiling to rub her back. For twenty or more years she relentlessly pumps out thirty thousand eggs each day. How often the king has to mate with her is a secret still held by the royal chamber, but somehow he manages. Gangs of workers queue at her rear end and carry this river of eggs away to dedicated nurseries, where they hatch directly into a nymphal form – translucent spun-glass miniatures of the workers.

The mounds were a story in themselves, providing homes – in their chimneys and chambers – for a great variety of creatures, from snakes and giant lizards to mongooses, porcupines and birds. I longed to get started. After the success of our earlier films we now had carte blanche to choose whatever subject we wanted, but when *Survival* and the US Network heard of my plans they went into shock. I was sent a ticket to New York and told that the suits at NBC wanted a serious talk with me. Days later I sat facing several prosperous-looking gentlemen at a vast Manhattan boardroom table that had once been a wonderful rainforest tree, host to monkeys and parrots. I was in alien territory.

'You've shown us underwater hippos, nesting hornbills, the wildebeest herds and ballooned over Kilimanjaro,' the suits said. 'These films were a great success, but we have to keep up the momentum, we need to follow them with something big and charismatic – like elephants or lions.'

I explained that all human life was there in those enigmatic mounds, that this was what I had set my heart on, and felt that I had earned that right, given my previous successes.

The suits looked pained: 'Alan, they are bugs, which is bad enough. But termites are worse than bugs. Millions of Americans own their own homes, they have mortgages and their homes are made of wood. In America, "termites" is the same kind of word as cancer.'

I started to explain that the film wouldn't just be about insects, there would be many other extraordinary creatures involved, such as . . . 'We will never get a sponsor. There's just no way we can place a prime-time special about termites. You need to rethink your plans, none of us wants to lose this slot.'

They simply would not take the time to hear about the aardvarks, cobras, barbets, mongooses and pangolins I would also be filming: they were looking at their watches and that pissed me off. I badly wanted to call their bluff and say, 'Just watch me,' but I wasn't certain enough of my case, so instead I answered, 'Well, I hear you, but I haven't let you down so far, so let's see what I come up with and then you can decide.' I left New York under a cloud, worried by what I had done, but now absolutely determined to follow my heart and make it work.

Back in Kenya, I spoke to Dr Jo Darlington, an entomologist who was studying termites in the dry country not far from where my father had taken that picture of me. Jo was a source of much useful information on termite society, the most interesting – and worrying – fact being that from the first blow of a spade on the mound, you have a society at war. The soldiers pour into the breach and major workers race to repair the damage, while others evacuate eggs and nymphs deeper into the mound. Open up a mound and you never see natural behaviour and of course if you don't open it you see nothing. Even if we were able to observe natural behaviour, filming it would be an even greater challenge, for the termites' soft bodies melt in the heat of photographic lights. (All this was, of course, before the advent of fibre optic tubes that can be eased into a mound to film without greatly disturbing the inhabitants.) Several scientists and film-makers told me that the film would be impossible to make, but I knew it was a riveting story and felt there had to be a way. Some ingenious friends at Oxford Scientific Films had solved the heat problem by shining a light through a ten-gallon glass demijohn of water, which focused to a bright but cool spot. That was fine in the laboratory, where they had filmed their insects, but was too unwieldy for the bush. Making further enquiries, I discovered from a doctor friend that a new light had just been developed in America for making surgical training films. Obviously you can't use hot lights on an opened body or you end up with a mixed grill, so these lights were bright but cool and I sent off for some to start experimenting.

Jo's nation at war simile was a more intractable problem and, after many attempts, it seemed that those who had said the film would be impossible to make had been right: all I had been able to film was desperate defence and rebuilding activity. When tens of thousands of termites come to the defence of an open mound there is an air of crisis that is compounded by the constant contact between frantic individuals, all passing messages of distress – I decided to try to minimise this troubled atmosphere by cutting out about a cubic foot of the mound, which I moved a short distance to a tent specially made of ground sheet material, which kept it dark, warm and humid. Using fine entomological tweezers, Joan and I painstakingly removed all the soldiers and the major workers who were the main responders to emergencies. Without them fussing about and spreading alarm, within a surprisingly short time the minor workers – the carers and cleaners in the society – instinctively got back to work. So with our new cool lights we were able to film the tiny freshly hatched translucent nymphs begging for food like puppies, along with the workers taking care of the fungus gardens, nurseries and all the other activities inside the mound. I also managed to open up the royal chamber and film the other-worldly life of that strange royal couple, where before there had been perpetual darkness. We became very attached to these extraordinary creatures, and after filming them in our special tent we would return the sections and watch the excited workers reintegrate them back into the mound and their society. We were getting there.

The point I had been trying to make to the suits in New York was that termites themselves are fascinating, but so too are the many creatures who move into the mounds with them. Warm and humid, the many tunnels and chimneys that make up the termites' air-conditioning system are a perfect refuge for snakes and lizards, who often lay their eggs in these natural incubators. The savannah monitor, a heavily built lizard with a massive head that stalks the bush high-legged in search of other reptiles, nestlings and eggs, will spend the whole of the dry season inside a mound in a 'summer sleep' called aestivation, the antithesis of winter hibernation. The spitting cobra is another common resident, and when these two meet the battle is fierce and usually fatal.

One day a four-foot lizard was looking out of the top of a mound, bright pale eyes scouting for anything edible and long blue-grey tongue periodically flickering out to taste the air. A cobra, almost six feet of shining gunmetal, wound its way slowly up the mound, searching the tunnels for prey, and came face to face with the monitor. What followed was almost robotic, as these two reptiles went about mortal combat as if sleepwalking. The monitor simply grabbed the cobra by the body and the two rolled down the mound and crashed to the ground in a cloud of dust. The monitor's way of dealing with large prey is to shake it violently to death, but in the fall it had lost its grip and the cobra now reared up and spat into the lizard's eyes. Most cobras have short hollow fangs, which inject their venom from the tip, like a syringe needle. The black necked or spitting cobra's fangs have their exit at the front of the fang, so that when the venom sacs are forcefully squeezed by its powerful jaw muscles the venom comes out in a forward-facing stream. They are incredibly accurate, can shoot up to ten feet and deliberately aim at an enemy's eyes. The venom causes great pain and temporary blindness, but the monitor lizard merely gave a couple of blinks of its translucent inner eyelids, then lunged forward to grab the snake halfway down its body. Rearing high on its front legs, it swung its head from side to side with tremendous power, beating the snake hard on the ground. The commotion attracted a pack of bright-eyed dwarf mongooses who shared the mound with the lizard. Alert and constantly active, these small predators were fascinated by the battle and the whole pack of twenty or so came to within ten feet of the action, taking turns to stand on their hind legs and churr their alarm calls. Now the cobra had got its fangs into the lizard's back, and its violent attempts to dislodge the snake showed that this was obviously painful. Monitors, however, are immune to the venom of most snakes and, apart from hurting locally, the bite had no other effect. The snake had been weakened by the battering and after another minute of being slammed to the ground it was dead. The mongooses were apparently cheering for the lizard and moved close to watch as, tired and breathing heavily, it walked to the cobra's head, picked it up and slowly swallowed it whole.

Inside the mound, one class of the society had been undergoing radical change. Growing to an inch long – the size of their king – they had sprouted two pairs of fragile wings and, unlike the millions of sterile soldiers and workers, they were fertile. Called alates, they are the next generation – princes and princesses – who must now await the rains. Sensing the barometric changes that warn of imminent rain, the workers open exit slits around the base of the mound and then, usually just after dark and the first heavy shower, soldiers fan out to protect the great outpouring of the alates. They slide out through the slits in an unbroken stream of tens of thousands, open their four diaphanous wings for the very first time and rise rapidly into the night sky. Sometimes they will climb to the top of the mound to take off there, cloaking the ten-foot chimney in a sheath of living silver as the moonlight catches their wings. Their faint rustlings provide a whispery soundtrack to a magical scene. Landing downwind of the mound, the females raise their trembling abdomens and emit a pheromone to attract the males. The pair then go off in tandem, stopping only to drop their wings, almost as if they wanted to slip into something more comfortable. They quickly dig a hole a couple of inches deep, where they will mate and start a new colony; the princess even brings some spores from her home mound to start the fungus gardens. To tide them over the next few weeks while they are getting set up, alates are packed with fat and protein, making them a rich source of food for a whole alphabet of creatures that come to feast – everything from hyena to praying mantis, and warthog to chameleon, as well as several tribes who capture, roast and eat these nutritious morsels. The Pokot people of northern Kenya even have a system of ownership of individual mounds, which are passed down the generations. At the beginning of the rains, the Pokot cover the base of their mound with cowskins, sealed with earth around the edges, leaving one opening in which they dig a foot-deep pit. When the alates emerge under the skins they head for the light, fall into the pit and are collected in handfuls. When roasted and winnowed to get rid of the wings they taste a bit like peanut butter and are considered a great treat.

Filming the termites took us two years. Everything came together in a dramatic ending when army ants raided a mound that had been broken open by an aardvark. There was an epic battle in which the golden termite soldiers were overwhelmed by the shiny black ants, who then poured into the mound and killed the queen. With no queen to add thirty thousand eggs a day to the society, the remaining termites soon died out and the colony fell silent. No longer guarded or maintained, the mound slowly began to erode. Like great sanding machines, elephants scratched their itches on it and wind and rain finally reduced it to a bare patch of earth. But one day a pair of alates would land on that patch, find each other, shed their wings and start to dig, and the whole cycle would start again. From the balloon I framed a tight close-up of a pair digging that slowly lifted away in a single shot to reveal their shed wings, the worn patch and finally a wide view of Africa from a thousand feet. It made a wonderful end-title shot, similar to the one I had envisaged with some daisies on an English lawn five years earlier. Some things are worth waiting for.

The film was what I'd always known it could be. Not just a unique look at the 'bugs' that build those mounds, but an in-depth ecological story that wove in the extraordinary range of creatures that come to the mounds for food or shelter. When I showed the finished film, *Castles of Clay*, to the suits at NBC they loved it and said, 'Why didn't you tell us it would be like this?' I tried to tell them that I'd tried to tell them, but again they were looking at their watches. The film went on to win many awards, including a prestigious Peabody Award, one from the British Association for the Advancement of Science, and was nominated for an Oscar.

Because of the many new things our films were bringing to light, one of the questions we were most often asked at that time was 'How the hell did you get?' followed by 'nesting hornbills', 'spitting cobras' and 'underwater hippos' – the list goes on. Exhausted after two years of working on termites, we decided it would be fun to make a less demanding film this time round, so with the working title of 'How the Hell?' we started on one showing how we had shot some of those key sequences. As Joan and I would have to be filmed, the cameraman we had worked with on *Safari by Balloon*, Martin Bell, joined us again.

It was meant to be a fairly easy quickie, but once I got into it I was driven to show not just how we had got some of our memorable wildlife sequences, but greatly to improve on them. Given this second chance, maybe I could even perfect them? I now wanted the cobra spitting its venom in ultra-slow motion. Although it looks very dangerous, if you know what you are doing a confrontation with a spitting cobra is easily choreographed. When approached, the snake rears up and looks straight at its target – your eyes. Stay where you are and it will do the same. Push your face closer and at a certain point the snake will open its mouth wide, give a powerful squeeze to the large venom sacks in its head and send a stream of blinding venom straight at your eyes. If, after that, you decide to push even closer, the snake will strike and bite, but move back a foot or two and you return to status quo ante – a stand-off. Indian snake charmers use this technique to keep their cobras in the reared up-posture. Joan was familiar with snakes and, with her specs for protection, was happy to be the target. With a special high-speed camera I was able to get spectacular shots of the silvery streams of back-lit venom as it flew towards her face. I was often censured for putting Joan in this position but together we had decided that it wasn't all that dangerous, and if done with no fuss by an unassuming woman it would be less theatrical than with a khaki-clad male. So it was, and proof came many years later with Steve Irwin and his macho mugging and posturing in a similar sequence.

The sequence of us refilming the termite hatch in a rainstorm was a comedy of drenched equipment and exploding lights, while the cheetah hunting and wildebeest river-crossing sequences were a great improvement on the originals. It was eight years since I had filmed the hornbills nesting. I was convinced I could now do a much better job and this time chose the more colourful yellow billed as the star. I decided to film them from George Adamson's camp at Kora, on the Tana river, where they were plentiful.

George, an ex-game warden, had, with his now ex-wife Joy, become famous for raising the lioness they had called Elsa, and had been avoiding that fame ever since. Camped out in this hot dry country for many years, he released a string of lions back into the wild with varying success. He lived inside a high wire-mesh enclosure that kept the lions away from a small group of thatched huts at the foot of Kora, a towering red rock several hundred feet high. Dress code for gents was shorts and sandals, and after a day of walking with his lions George was always ready for a drink. We would move camp chairs close to the wire and settle down with a whisky to watch the changing colours on Kora rock. Every evening, without fail, his brother Terence, a lifelong teetotaller, would place his chair just downwind from George. Once we were all comfortable, Terence would deliver an explosive snort of disapproval and, cursing the 'stink of alcohol', move his chair upwind of George. After observing this daily ritual peace would return, and we sipped quietly as various lions ghosted in from the darkness and came to the wire for titbits of camel meat.

Terence was a very knowledgeable botanist and skilled at making roads through the bush; and though they constantly bitched at each other there was obviously a powerful bond between these two weather-beaten characters. But Terence shared none of George's feelings for the lions, he really didn't like them – and the lions seemed to recognise this and reciprocate in kind. Another camp chair was occupied by Tony Fitzjohn, George's young assistant, a wild, randy fellow with London working-class origins like myself – I called him the bush-wise barrow boy – and we got on well. He had become adept and fearless in handling the lions, and indispensable in nursing George's battered Land Rover and a tractor that I had persuaded Dr Grzimek to donate. During the three months we spent at Kora we were often able to accompany George and his lions on their daily walk down to the Tana river. It was amazing having five or six great golden cats padding alongside, playing with sticks and chasing each other around. George was quietly confident that we were safe, but I have to say that I felt happier being near wild lions than these big cats that had lost all fear of man.

We had torrential rains while we were there, which brought huge quantities of insects, and the hornbill we were filming raised three broods. Each time the female moved back into the nest soon after the chicks had left. I was filming inside the nest as before, but this time I had a better understanding of the breeding behaviour, improved equipment and three chances to get it right, so we got some really superb footage. Working on foot knowing George's lions were somewhere nearby we were always alert. We paid particular attention to Shade, a young male who often stalked and charged us if we got too close to the wire in the evenings. We were right to be concerned, for early one morning Joan and I were having coffee when a scream brought us racing to the gates of the compound. Terence had been outside the fence burning some old thatch material and now he was down and a lion had his face in its mouth. We ran towards them yelling loudly and some of George's workers threw rocks at the lion. Sure enough, it was Shade, who dropped Terence and backed off, growling, while we dragged Terence back inside the wire enclosure. He was covered in blood, his teeth visible through a gash in his cheek where a canine had narrowly missed his eye, and there were deep holes in his neck that had somehow avoided his windpipe and carotid. We disinfected the wounds, bound his head lightly and rushed him to my plane. He was shivering with shock, so we wrapped him in a blanket and Joan held him tight all the way to Nairobi. Terence was a tough old bird, made a remarkable recovery, and later said that his main memories of the event were being cuddled for an hour by Joan and having his first hot bath for twenty years – all on the same day as, incidentally, being mauled by a lion.

George hardly ever left Kora, so over the following years we visited periodically, often with a projector to show him our latest film. On one of those trips we flew him north to Shaba, on the Uaso Nyiro river, where his ex-wife Joy was following her successful rehabilitation of a cheetah by doing the same with a leopard. None of us could understand how Joy was able to work with a volatile creature like a leopard and not get chewed up. She had a home close to us at Naivasha and whenever she came to visit our staff would hurry to close away our more timid animals before Joy's exuberant greeting could send them slamming into the fence in retreat. She would advance far too rapidly, holding out her arms and loudly calling, 'Come, come, come' – the subtext being: 'I'm Joy Adamson, you must know me, trust me, love me. Come, come, come.' I have never known anyone with fewer social skills when dealing with animals. At her approach our antelope would bolt in terror and even the aardvark, fairly short on social skills itself, would dive down the nearest hole. I was far from alone in this view of Joy, yet there is no denying her great success in working with lions, cheetahs and leopards: gaining their trust and affection while seeing them return successfully to the wild. She and George enjoyed seeing each other now and then, as long as the visit was short, and on that trip to Shaba Joan took a wonderful picture of them together in that wild setting that I still have on my desk. Little did we realise that this would be their last time together. A few months later, in 1980, Joy was found dead near her camp, and it was reported that she had been killed by a lion. It was generally felt that this was an appropriately newsworthy end that she would probably have found fitting – as long as it wasn't one of George's lions! When it later turned out that she had been murdered by a disgruntled employee, I was saddened by the banal manner in which she had left this land, which she had made a much more interesting place. Nine years later, George was gunned down by Somali bandits as he charged to the rescue of some tourists ambushed near his camp at Kora, going out in one of the ways he would have chosen, fighting for the things he so fiercely loved. The roaring of a lion still brings back memories of those two unique individuals who led such extraordinary lives, the like of which can never be repeated and in which we were lucky to share.

The hornbill sequences we filmed at Kora had wrapped up shooting for the 'How the Hell' film, which we later renamed *Two in the Bush* – and which the Americans, in order not to upset the Kraft cheese people with the possible sexual connotation of the word bush – called *Lights, Action, Africa!*.

The prolonged rains at Kora that year didn't relent and what was normally a red, sand-blasted landscape of gaunt, leafless trees, was now clothed in greens of many shades. Grasses and flowers from bulbs, corms and seed that had been dormant for decades sprang up to cover the ground. The bush was aflame with the clashing colours that only Africa can wear – deep purple thunbergia, orange ruttya and yellow bidens daisies – all fighting for space amid dense convulvulus creepers, whose large white flowers covered acres of bush like snowdrifts. Elephants that had become accustomed to chewing dry sticks were now up to their eyes in flowering fodder, and the bush rang with the breeding calls of guinea fowl, hornbills and bustards. At dark the soundtrack changed, and from every waterhole, stream and puddle came the croaks, barks, peeps, rattles and groans of a great amphibian ensemble. Rather than going back to Naivasha to start editing the 'How the Hell' material, I decided we should stay at Kora and take advantage of the situation. For over a decade, whenever I saw the opportunity, I had been gathering footage for a film about the seasons in Africa. I had dramatic scenes of drought in the can, and had filmed the strategies of several frog species as they prepared to face the dry season and months of dehydrating heat. Now I was able to film the varied and exuberant ways these species went about maximising the – usually – short wet season to ensure the next generation.

One of my subjects was the foam frog, a pale, skinny creature just two inches long that spends the dry season clinging to a branch, all four legs tightly tucked in to expose the minimum of skin. That skin turns chalk-white to reflect the heat and is protected by a layer of mucus that hardens and is as effective as a plastic bag in keeping the frog from dehydrating. With the first rains these shrivelled, ghostly creatures quickly fatten on the plentiful insects, then, bursting with energy, begin their outrageous breeding behaviour. When she is ready the female positions herself above water, where she is soon joined by any number of amorous males. Along with the eggs she produces a quantity of mucus, which she and the males beat into a glistening white foam by vigorously kicking their long hind legs. Other males and females are attracted to the action, and come crawling through the branches until up to twenty or more jockey for position in a writhing, slippery mass of foam and muscular pumping thighs. Think hot-tub orgy meets the finishing stretch of the Tour de France and a funnier scene in nature is hard to imagine. Yet questions remain. Did this system evolve as a cooperative effort because a lone pair would not be able to produce sufficient foam for a large enough nest to protect the eggs? Or is it another case of the scrambling sperm competition that characterises the mating of many frogs? The research is not in yet, but meanwhile it looks a hell of a lot of fun. The foam forms a football-sized nest hanging above the pool, making the waterholes in the dry country appear to be decorated with candy floss. The outer skin soon hardens, effectively waterproofing the nest. The eggs inside develop into tadpoles which, when ready to break free, thrash their tails to liquefy the foam and fall through into the water below.

The foam frogs are just one of the creatures brought alive by the changing seasons, for the rain-softened earth now produced a mixed crop that had been buried in the hard red ground for many months. The giant land snails that had been sealed tightly inside their shells, send long-stalked periscopic eyes to the surface for a look around, before hauling themselves out of the mud like surfacing submarines; and a bullfrog the size of a grapefruit crawls to the surface, reaching back for the protective cocoon that has kept it moist for months, pulling it over its head like a rather grubby, translucent nightie, and promptly eating it: a first meal after months of summer sleep. The tiny eggs of the killifish, that have lain dormant in the dry earth, hatch and rapidly grow into colourful adults who mature and mate in a week. As their pools begin to shrink again the brilliant red-and-blue males frenetically display to females, who push their eggs deep into the mud as they are fertilised. The killifish die when the pools dry out, having lived perhaps no more than three weeks. But the next generation is there, pinhead-sized eggs, waiting – for years if necessary – for the next rains.

The lungfish, often found in those same pools, is an ancient creature able to breathe air and move overland using its lobed fins. When its home pool starts to dry out it can walk a short distance to a deeper pool, but eventually it burrows down into the mud, folds itself double with head and tail facing up, and seals itself inside a cocoon of mucus that will keep it moist and alive. By blowing bubbles on its way down through the mud it creates a pipe up to the surface through which it can take in air. Tightly encased in the dried mud, its metabolism slows down to just two per cent of its normal rate, and there it waits, scarcely breathing, immobile and bent double in its earthen sarcophagus for months or even years until the climate above ground is right for it to re-emerge. I find it remarkable that when water finally floods its burrow the lungfish, after such prolonged inactivity, still has the strength and muscle tone to wriggle its way through the thick mud to the surface and swim away. Perhaps this is something for long-haul astronauts to look into? In any case, for the film I was able slowly to slice away the hardened earth to get shots of the fish dormant in its silvery sleeping bag and later bursting out of the sticky mud.

All these sequences of rebirth were a telling counterpoint to the desperate scenes I had filmed in the terrible droughts of earlier years. One year Lake Katavi in southern Tanzania had dried out so thoroughly that only a large puddle was left in the centre of its deeply fissured bed. Stranded in that wallow were hundreds of hippos and many thousands of huge catfish. Even in a shallow puddle hippos can flail their tails to splash water on to their backs, but in the thick mud of that giant wallow they were unable to cool themselves and suffered terribly from the heat. Among their tightly packed bodies great numbers of three-foot-long catfish thrashed around, trying to keep out of the sun. Able to breathe air, they slithered and slid over the hippos' jostling backs, squeezing between their great heads, desperately searching for space and moisture. To film these scenes Joan and I had flown the six hundred miles from Nairobi, landing on the dried lake bed not far from the wallow and spending several nights sleeping out in hammocks hung under the plane's wings.

Hippos normally come out of the water at night to graze, and those that were strong enough now did so, but did not have the energy to walk the mile to the shore where there might be grass, so they simply stood quietly in the moonlight. Soon they were joined by a herd of gaunt, exhausted buffalo, who pushed their noses among the catfish and tried to drink from the mud. Then they too stood motionless under the moon, as if in a grim, silent anteroom for death. Then, on our fourth day heavy rain fell on

the nearby hills. I longed to stay and film the wallow filling up
and the hippos being washed clean, but I also knew that even
just one shower could make the black lake bed so sticky that
our plane would be unable to take off. More rain clouds were
building, and we could not afford to wait any longer. Our final
shot was of the wallow and buffalo with a faint rainbow in the
distance. As we flew away we could see a flood coming down
out of the hills that would reach the lake in an hour.

In that same year in Tsavo Park we had come upon a
distressing scene of a small group of elephants with a two-year-
old calf that had collapsed from hunger and exhaustion. A cow
stood on either side of the baby trying to urge it to its feet. The
distraught adults tried everything, pushing a trunk or a tusk
under the youngster and lifting it upright, where it hung like
a rag doll, only to slip back down; its little trunk, normally so
mobile and expressive, dangling limply in the dust. They braced
it between their legs, but it was obviously never going to stand
unaided again and we watched in anguish as the two females
persevered, all the while giving low moans of encouragement
and frustration. They kept up their efforts for eight hours, while
the rest of the herd stayed nearby, picking at the dry branches
of dying trees or standing in deep, patient silence.

Then, at last, the matriarch, one of the two solicitous cows,
stepped away from the calf and slowly began to lead her desperately thirsty family away. The mother moved as if to follow
the herd but after twenty yards she stopped and stood motionless for several minutes. Then she made the decision I would
expect of an elephant. She turned back to her calf, stood over
it and finally laid her trunk across the baby's back in a caress.
An hour later we left her there as darkness fell, monumentally
immobile, mourning and softly rumbling to keep contact with
her distant family.

9

The Big Broccoli

In the late seventies and early eighties Tanzania was going through a xenophobic period, with newspapers and radio endlessly telling the citizens never to trust foreigners. The borders were closed without warning, and many aircraft and hundreds of Kenya-registered tourist vehicles, including three belonging to our safari company, were parked in ranks on a Tanzanian airfield, confiscated for the crime of being in the country when the borders closed. On the Serengeti the tracks were grassing over and it was as empty of people as it had been on my first visits in the fifties. Almost uniquely, because of my long association with the Serengeti, we had police permits from both Kenya and Tanzania allowing us to pass back and forth. Kenya was agog because Robert Redford and Meryl Streep were filming *Out of Africa* at the time, and I was contacted by Redford, who had heard that I was probably the only person who could take him to the Serengeti. So we flew down along the Rrift Valley in my old Cessna 180, past the volcano Ol Doinyo Lengai and the crater highlands – all of interest to him since he had studied geology at college. Once over the Serengeti, I decided to land among those great tumbled rocks, the Moru Kopjes, to have a picnic and show him a rock gong – a huge flake of granite pocked with holes where it had been struck with round stones to produce a series of lovely bell-like notes. Hit by whom, and when, no one could say, but it was said to be one of the earliest musical instruments.

The grass was long and I knew there were rocks hidden there. If I dinged one it could be very embarrassing, for halfway through shooting a film Redford was certainly not meant to be flying about in a small single-engined private plane, let alone landing off-field. No doubt the anxieties got the better of me because sure enough, swerving to avoid a rock, I did a ground loop. Most small aircraft have two main wheels and one under the nose, a more stable configuration than the 180, which is known as a tail-dragger, with its third small wheel at the back below the tail. The saying goes that there are only two types of tail-dragger pilots: those who have ground-looped, and those who are going to. It happens when the aircraft is not kept dead straight on landing – usually in a cross-wind – and the tail whips round violently as if on casters. Do this going too fast and you can tear off the little tail wheel, or dig a wing tip into the ground, terminating your flying right there. On this occasion we were lucky and spun round in a cloud of dust, ending up facing the way we had come.

'Wow,' said Redford. 'That's what we call a Talmantz turn.'

'Oh, really,' I croaked.

'Yeah, I did a movie once called *The Great Waldo Pepper* about barnstorming pilots and we had these great stunt guys from Talmantz Aviation who would do that to make a really short landing.'

Here was a chance to recover my shaken self-image. 'Well, over here it's called the Root Loop,' I said. 'And you're right, it works really well.' He chuckled, but I'm not sure he actually believed me.

We stayed in the little house I was renting from the deserted Research Station and the next day took a long walk against the tide of the dense herds of trekking wildebeest, which parted just ahead of us and then re-formed behind so that we walked inside a small, moving circle among the flowing, grunting animals. When the time came to head back, Robert asked if we could fly via the Mara Game Reserve, which is always crowded with tourists, even more so then because Tanzania was closed to visitors. At one point I flew towards a group of tourist vehicles, over twenty of them, cameras clicking, in a tight circle round a small mound on which sat a fine golden-maned lion. As we circled above them Redford looked down and I could feel the real sadness in his voice as he said, 'I know exactly how you feel, boy.'

With each production, when the filming in the field was over, we would move back to Naivasha, where I would spend a month assembling an almost complete edit and writing up the commentary, sitting below a sign cautioning me to 'ESCHEW OBFUSCATION'. Joan quickly settled back into home life and enjoyed cooking in a proper kitchen, visits from friends, working in the garden and caring for her wild pets. Our neighbour Jack Block would come down with interesting guests from his hotels almost every weekend, bringing them over to see our animals and inviting us to lunch. Joan loved meeting artists and writers, and other people with good conversation and stories, so by the time I was ready to leave for London for the rest of post-production she was comfortably settled and didn't want to join me. Not only was she happy at home, but she really didn't like England, could do without the shows and museums – there was no way I could change her mind. I would then head to London for the six weeks or so it would take to finish the film – the final edit, the soundtracks, titles, colour correcting, the recording of the narration and the music, dubbing, and then the long, exciting day of the final mix when it all came together and two years of work was nestled into a can the size of a small dustbin lid.

In London I was *Survival*'s poster boy. I was handed the keys to the flat, the car, tickets to *Cats* and Wimbledon – all the crap that lures a man into thinking he is more than he knows he is deep down. *Survival*'s offices and cutting rooms were in Park Lane and there would be picnics in Hyde Park or well-lubricated lunches at Fino's wine bar next door. I was eating, drinking and partying far too much, and although I never failed to deliver a good, award-winning film on time and on budget, I could feel I was getting lost.

For all my worldly success I felt a failure in what was becoming a more and more important issue. I was desperate to have children. Joan's inability to conceive was a deep sorrow for her, but one she buried, perhaps because of the strange way she had been raised. When she was small, it had been some supposed expert's theory that if a baby was comforted every time it cried it would grow into an adult who would cry whenever it wanted help from others. Over her mother's objections Joan's father had insisted that when she cried she should not be attended to or comforted – that way she would grow up self-sufficient. As her baby cries had never been answered, Joan had grown into someone who would never call for help; she was unable to discuss how she felt about our situation and instead gave her love to the wild animals she raised. This seemed to assuage her needs, but though I loved them too, wrestling a ratel or hugging a hyena provided small solace, and at times my loneliness was a bone-deep ache.

I had been chased by enough nursing lionesses with cubs to be familiar with the fierce maternal instinct among female animals, but had no idea it could be so powerful for a man. Africa and her animals were no longer enough. Seeing kids sitting on a Land Rover roof or putting up their little tents, I felt desolate and incomplete, my life unlived. I became borderline suicidal and, three times in as many months, purely through neglect, I ran out of fuel in the plane, making two successful forced landings and one not so pretty. In my work I had many times risked my life, but it had always been precious to me; now it seemed to have little value. I drove my Range Rover like a maniac, Walkman headphones on full volume, tears streaming down my face, bellowing along with Kris Kristofferson's gravelly ode to loneliness, 'In a park I saw a daddy, with a

Elephants drinking at Mzima Springs while I get
used to being in the water with the hippos

Undisturbed by my presence, a hippo 'moon-walks' past

Me, aged seventeen, displaying my snakes at the Nairobi Agricultural Show

A little later, with a Gaboon viper, an A-list venomous snake

The morning after the puff-adder bite. The next day the blister had filled my hand

Recommended therapy after having passed through a mouth similar to the one on the right...

The old way of crossing rivers in the Congo: 20 tonnes of illegal timber belonging to a government official come a cropper

The new way of crossing rivers in the Congo: now that the bridges have gone

My six-wheel drive Pinzgauer in its element

The male Congo peacock, the celebrity bird that after all that effort was ultimately rather boring

The aquatic genet

Me with (the incredibly rare and shy) aquatic genet

Holding one of Africa's tiniest primates, Demidoff's bushbaby

Huge herds of elephants once roamed East Africa's national parks. Garamba, in the Congo, was the last place where herds of many hundreds could be seen

This is how so many of them ended up

With Dr Bernhard Grzimek
in the Serengeti in 1963

Joy and George Adamson at Shaba, one of
the last pictures of them together

At the royal performance of our
Galapagos film at the Royal
Festival Hall in London

Jackie Kennedy after our
balloon crash-landing

The three biggest lions in the Ngorongoro Crater climbed to the rim and spent many hours in front of the grave of Bernhard and Michael Grzimek

Today there must be literally millions of images of gorillas. Back in 1961 this was one of the best from a tiny selection

Myles and pet rock hyraxes
Pia, Pumpkin and Potato

Rory and Gema, the Sykes monkey
that the boys called their 'brother
from another mother'

Rory, Fran, Myles and Me

laughing little girl that he was swingin'.' I remember once sitting in my car, alone out on the plains, wondering why I was feeling bereft at the sight of a whistling thorn. These short trees are covered with large thorn-bearing galls, full of holes over which the wind whistles mournfully and which, when painted gold and silver, make a lovely traditional Kenyan Christmas tree. I realised it was mid-December, and I was in tears because I had no children, no reason to decorate that tree.

Frightened by what was happening to me, I tried to bury my feelings with work. God, what a familiar story! I suddenly decided I wanted to make a series of six one-hour films about the Serengeti ecosystem. The outlines I wrote for *Survival* were readily accepted, but it was a big project and I would need to employ at least two top camera people to help, something I had never done before. Except for the company of my soulmate Joan, I was very much a loner when working and knew I would not be good with employees, but I signed on for the project anyway. I was also planning a series of films in the rainforests of the Congo, again for *Survival*, to look for the okapi, the Congo peacock and the other mysterious creatures that had never been filmed. This project would be a huge logistical undertaking with a good chance of failure due to the endlessly volatile political situation in the Congo. Again, I knew I would need another cameraman and various assistants. Again I signed on, for four one-hour films. With very little thought I had committed to ten hours, in two countries, six hundred miles apart, and neither *Survival* nor I had ever made a proper series before. I was riding the wave of my successes to date – on the surface confident to the point of arrogance, but deep down trying to bury my churning emotions. No one in *Survival* would tell me what must have been blindingly obvious to them – that I was biting off a whole lot more than I could chew.

When I got back to Kenya I was a mess. A couple of days later there was a big wedding in our Naivasha garden at which I got hog-snarling, boorishly drunk and started to become entangled with a lady who appeared already to have me in her sights. Jenny was a tall, athletic potter, with an infectious laugh and a head of curly hair like a golden chrysanthemum. Born in London, around the same age as me, she had spent most of her life in Kenya and was in a marriage she was seeking to leave. We soon started an affair that shocked and angered Joan and my friends, and added to my sense of disorientation. I was rudderless, understanding nothing. At first Joan and I struggled on as though it weren't happening, with both of us hoping that everything we had built together would win the day.

We flew to the Congo, where we spent a couple of weeks negotiating with the authorities and visiting the Ituri forest, working out a plan for making a filming base there. We staggered along in this messy, unresolved way for a couple of years, and as impetuous passion and righteous outrage cooled, there seemed at last to be light at the end of this painful tunnel. By this stage Jenny and I realised it wasn't going to work out between us, and I had found her a house – and built her a pottery – on a friend's ranch about fifty miles from Nairobi, with wide views out towards Kilimanjaro.

I went back to England to try to persuade a couple of promising young film-makers, Mark Deeble and Vicky Stone, to come and work for me on the Serengeti series, when I received a telegram from Joan: the volcano Nyiragongo was erupting, she wrote, did I want her to get flight clearance for the Congo? Oh yes I did. We had spent a wonderful night on the rim of that volcano twenty years earlier, watching the waterfalls of melting hail hanging from the walls of the fire-filled crater. Now a side vent was throwing fountains of lava hundreds of feet into the air. This felt like the catalyst Joan and I needed finally to get us properly back together again, and I left Mark and Vicky to decide their futures and headed back to Nairobi, Joan and the Congo. We spent a week there, camped just out of range of the molten lava bombs, some of which landed uncomfortably close, flattening with a whoomph into glowing orange cowpats. Rivers of lava raced down the slopes, then slowed to walking speed as they met the forest, setting ferns ablaze and pushing down trees that toppled slowly, bursting into flames as they met the molten flow, the loud squeaks, whistles, pops and cracks as the wet wood exploded providing a dramatic soundtrack. At night, thousands of insects were attracted to the two-hundred-foot-high fountains of lava, and pennant-winged nightjars

hawked in silhouette against the orange sky, each wing trailing a single eighteen-inch-long feather. As we made our way back home we made more plans for the Congo project and our future together. It was great to be back in harness again.

Not long after returning to Nairobi I got a call from Jenny, who had just come back from a trip to England. She asked if I could drive her home, as she was recovering from an operation and felt she shouldn't drive. As we were no longer together, this seemed like an unfair demand, but she sounded desperate so I went to collect her.

'What's wrong? You look exhausted,' I said the moment I saw her.

'I'll tell you in the car,' she answered, then explained, 'It wasn't really an operation. I had a check-up and the doctor was concerned about my white-cell count, so he drilled a hole into my hip to get a bone marrow sample. Walking is still very painful.' My own marrow went cold at these words. I seemed to remember that this was part of the diagnosis for . . . I stopped thinking, went fearfully quiet and waited for her to go on. 'I've got leukaemia and if I'm lucky I have two years to live.'

I don't know how to describe the effect those words had on me. Anguish and despair, mixed with unfocused rage, and uncertainty about what this meant for all our futures. I drove on autopilot and dropped Jenny at her house. It was a perfect Kenyan day, with Kilimanjaro floating in a cornflower sky. My life was falling apart. I could not walk away now and after long and agonising discussions, in a wrenching attempt to make the best of the situation, Joan and I agreed that we could wait two years to get back together again, and that I should stay with Jenny until the end. If you want to make God laugh, tell him your plans.

I moved to Ulu to be with Jenny, but now I had to knuckle down to my filming commitments and, given the intense emotions that were churning around all of us, I knew it would be impossible to have even a professional relationship with Joan. I was going to miss her help sorely on such a big project, but I had to get started in the Congo. Growing up in Kenya, the Congo had always been our Wild West – a bigger, more dangerous frontier land of endless forests, huge rivers, volcanoes and mountains, wilder peoples and fewer laws. Now, under the despotic President Mobutu, the country and even the vast River Congo had been renamed Zaire. I had already made several visits, including the one working with the gorillas, but had always wanted to spend proper time there and I was now raring to go.

The films would be going to *Survival*, Grzimek and the *National Geographic*, but I didn't know a great deal about the troubled country I was headed for, so the outlines for the films were not yet very well defined and it was difficult even to guess at what I might be able to shoot there. Basically, I wanted to follow in the footsteps of Chapin, Lang and Cordier, and to find the Congo peacock, the okapi and the aquatic genet, but it seemed absurd to say that. By now, fortunately, I was trusted well enough to be able to be vague, and just promise I would be making one film each about the Virunga and Garamba National Parks, and one about okapi and the Mbuti pygmies in the Ituri forest. While I worked on those I would collect, over the years, sequences for a fourth film, which would take in all the most rare and virtually unknown creatures of the rainforest. I listed lots of these by their Latin names, which so thoroughly confused the suits that I was left alone to get on with it.

I was in the process of getting set up in Goma, on the shores of Lake Kivu, when I got a call asking if I would shoot some footage for *Gorillas in the Mist*, the film that Michael Apted was making about the life of Dian Fossey, who had been murdered two years previously. He wanted some shots of a silverback charging the camera as Dian had herself been charged in her early days. It was a bit of a diversion from my project, but I felt it would be a good memorial to Dian, so I agreed to help. The gorillas in Rwanda were very relaxed and accustomed to tourists, and I joked that they just sat around all day reminiscing about the time they met David Attenborough. They were far too relaxed with humans to charge and we certainly would not want to provoke them into doing so – but in Kahuzi Biega Park in Zaire I knew of just the right animal: a silverback called Mushamuka. He had been known since the early seventies, but the Park Warden told me that he'd become morose and aggressive, and for some time had been out of bounds to visitors. We agreed that this was the silverback I should visit and I would be guaranteed a charge.

With my assistant Bruce Davidson and a couple of guides, I set out to meet Mushamuka. I had been given a body mount for the camera, which meant it was permanently positioned in front of my face. I knew this would be seen as threatening body language to the gorilla and had discussed with Bruce how we could operate with the camera on a short monopod instead, so that I would be in a more submissive posture. But we wanted to test it out and headed off to find Mushamuka. We found him eating reeds in a swamp with the rest of his group. It was a bad start. Unable to move quickly in the water, he felt insecure and showed it. We crept out of his line of sight and sat quietly, giving them plenty of time to move off. After an hour we started to follow and soon found a young male sitting alone on the trail, nervous but curious. We approached slowly, and he let us get close before suddenly his nerve broke and he raced off, making a loud alarm call. Hearing that cry, Mushamuka reacted instantly.

A gorilla's charge is usually nearly all bluff. Standing up to show his frightening size, he beats his chest, gives a bone-chilling, roaring scream, showing large, blackened teeth, and makes a short, threatening rush. It may indeed be mostly bluff, but believe me, it works and has powerful diuretic qualities. Mushamuka wasn't bluffing. He came out of the forest like a bloody great Doberman on steroids, silently racing towards us on all fours, his huge shaggy arms reaching forward, powering a flat-out run. I had the wrong lens on but started filming anyway and the viewfinder was swiftly filled with black. The next thing I knew I was held in a powerful grip by my left leg and waist, lifted into the air and flung down hard. Then Mushamuka was gone, disappearing into the forest to grunts and screams from his family. He had picked me up the way a hungry man might lift a roast chicken and taken a bite out of my thigh. He hadn't wanted to kill me, just to tell me to push off, and I was thankful he had used so

little of his awesome power. I was helped to my feet and was examining my waist, where I felt sure I had been bitten, but it was only where his thick fingers had dug into that ticklish spot.

As with the hippo attack, I had not felt any pain at the actual site of the bite and hadn't noticed a large flap of flesh hanging just inches above my kneecap. A bit lower and I would have been crippled. Bruce tore up his shirt and bandaged the flap back in place, and we started on the two-hour walk back to the road. By now the muscles in my thigh were twitching uncontrollably and cramping up, and several times we had to run across areas that were carpeted with raiding army ants, but finally we made it to the car and drove to the airfield at Bukavu. There, claiming that I must have killed the gorilla in retaliation – *C'est naturelle, non?* – I was interrogated by police for an hour before they realised I was not going to bribe them and I was soon on a flight to Goma. Now the real pain started, not least when the doctor in Goma poured neat iodine into the wound, destroying a lot of tissue in the process. Then came another drive, this time into Rwanda to catch a plane to Nairobi where, some thirty hours after being bitten, I was once again in the pragmatic care of my good friend Imre Loeffler.

'So, vot vud one of your animals do ven it is vounded?'

'Well, I guess it would find a safe place in the shade, lick its wound and rest.'

'Zat is exactly vot I vont you to do . . . don't vorry about ze licking, zat is optional.'

So, as I later told him, 'Zat is exactly vot I did – and it verked!'

Warner Brothers got in a bit of a state about the incident and we were told to keep it very quiet. Sigourney Weaver had agreed to work with the gorillas in Rwanda, which were well accustomed to tourists, but Warner were worried that she might back out if she heard that the most experienced guy on board had been bitten on Day One. Three weeks and some plastic surgery later, undaunted – well, quite a bit daunted, actually – I was ready to go back to Mushamuka and try out the submissive posture. I reintroduced myself to him slowly over several days, and joked that once I'd shown him the script he was very helpful. In fact, he turned out to be a very interesting – if somewhat volatile – character. On many days he was playful and seemed to enjoy waiting and watching with interest as Bruce and I moved slowly towards him. Then he would rush at us with a full-on bluff charge with all the trimmings, sometimes whacking me in passing, or sending one of us flying with a shove of his massive shoulder. We rarely felt in any serious danger, though, and he never again adopted the Doberman position. He didn't get upset on those days, and although his behaviour looked impressive, it felt no more threatening than rough play. At other times, however, he was obviously in one of the dark moods that had given him his bad reputation, greeting us with a hostile roar and a 'make my day' aggressive posture. On these occasions we would slink away quietly without disturbing him. In a couple of weeks I was able to give the director what he had asked for, a selection of impressive charges. One of them being the bill I sent to Warner Brothers.

I also needed gorilla sequences for my Zaire project, so we moved from Kahuzi to a hut on the Virunga volcanoes. As a result of the visit Dr Grzimek and I had made together years earlier, the Frankfurt Zoological Society had been paying the park guards' salaries, and financing an operation to habituate the gorillas to humans, which would open up gorilla tourism in Zaire and a revenue stream for the parks. Tourism had yet to be established, so we had a unique opportunity: gorillas tolerant of close approach, but no tourists or the rules and time constraints that come with them. Bruce and I took full advantage of the situation and spent weeks with the animals. We were with them for hours at a time and a couple of times I watched them until they went to bed, before bedding down myself close by.

During that wonderful time I saw several memorable scenes

that confirmed my views about the gentle nature of these great apes. One day, moving slowly towards a group that were gathered around peering intently at something in the vegetation, I saw they had found a newborn duiker antelope that lay curled among the ferns. The silverback towered over the cat-sized fawn, slowly lowering his massive hand, palm up, and tentatively touching its coat with his knuckles. On contact he jerked his hand away and his wide-eyed group, crowding in curiously to watch, jumped back, muttering nervously. He reached

forward again, but when the fawn suddenly raised its head his nerve broke, and he and the audience trooped off.

Another time we got a wonderful sequence of a large group of gorillas bunched together peering at a foot-long emerald-and-orange three-horned chameleon. One immature ape, braver than the rest, pulled gently on its tail, enough to stop it walking off for a while but no harder. Adults and young all watched intently from a few feet away: the youngsters huddled close to their mothers for safety, their lips puckered, their wide, shining eyes totally transfixed. A couple moved in for a closer look and one poked the chameleon gingerly with a sausage-sized finger as it crawled off. Chimpanzees would probably have torn the reptile apart and fought over the spoils; the gorillas quietly watched it walk away. I've always felt that gorillas – powerful but calm and gentle – are what most humans would like to be. Chimps – more inventive, but rowdy, combative and murderous – are, I'm afraid, more what we are.

For three days we also followed a female with an infant that had died at birth. She would hold it to her breast, looking down to see if it had found her swollen teats, moving it to a better position, lifting the little head that fell limply to one side, all the time murmuring soft encouraging sounds. Whenever the silverback moved on she followed on three legs, holding the baby against her belly, but on the third day, with the baby now smelling strongly, she held back, touched it for one last time and slowly followed the group, her sense of loss palpable.

Among the old lava flows around the base of the volcanoes were spots that looked markedly different from the air. The palm trees were sickly, the grass yellowed and scattered with bleached bones. These areas are called *mazukus*, carbon dioxide seeps, where gas leaks from the ground and collects in hollows creating a deadly trap. Carbon dioxide is not poisonous, but can be lethal. Breathe some in and the lack of oxygen makes you gasp for more air, another breath or two and you lose consciousness, any more and you die. When I went in on foot to investigate I found, scattered among the anaemic plants, the bones and bodies of snakes and lizards, mice and monkeys, bats, birds, butterflies and buffaloes. All of them had flown, walked or crawled through the layer of gas – just three feet deep – and got no further. Even elephants have died in some *mazukus*, breathing as they do through their trunks, which are near the ground. The gas comes out of the earth warm, so I could feel its level on my thighs and could move around filming in safety. But I did need to get down for low-angle shots and close-ups. I was surrounded by evidence of how easily I could be overcome, so I stationed an assistant on higher ground nearby to watch over me.

The residents of nearby Goma, on the shores of Lake Kivu, know all about carbon dioxide. It bubbles up in the lake and, on calm days, forms a layer just inches above the surface – right where a swimmer takes a breath. Wise locals test the air just above the surface with a Bic lighter. No flame, no swim. Dogs die on lakeside lawns because they get out of the wind by lying in a sheltered hollow where they go to sleep, the wind drops and the hollow fills with gas. Where the road to Tongo takes a dip and fills on calm nights with what looks like mist in the headlights, residents know to take it at speed or the CO_2 will kill their engine. When filming the eruptions and lava flows from the plane, every blast of hot, smoky air that rocked the machine reminded me of that engine-killing power, and whitened my knuckles on the controls.

The victims I filmed in the *mazukus* ranged in age from the weathered bones of hippos and a chimpanzee that had died many years ago, to mummified bats and birds, and the bodies of a kingfisher and puff adder that were just days old. The smell of the dead snake had attracted a four-foot monitor lizard, which had approached along a lava ridge and climbed down for a feast. Almost immediately he had shown signs of distress, opening his mouth wide and desperately gulping for air, his pinkish tongue turning blue from anoxia. I thought he was a goner, but he turned away and, wobbling on weakened legs, tried to climb out of the hollow. He didn't have the strength to make it, but had obviously got his head just high enough to breathe air. He lay there for ten minutes, regaining his strength, then hauled himself up the bank and crawled away. Scavengers coming to feed on carcasses like this made up many of the victims, adding the bodies of vultures, jackals and hyenas to the strange and spooky cemeteries.

When I'd finished filming around the volcanoes the plan was to move north to the rainforest, but first I was off to America to join Jenny for various appointments with top leukaemia experts. The news was discouraging: she needed a bone marrow transplant, but a worldwide computer search had failed to find a suitable donor. We drove up the California coast on the spectacular Highway One, the alternative medicine and lifestyle capital of the world, where we spent some time meditating with a nice man with a ponytail, whose name escapes me. It was a dispiriting trip, but Jenny still had her sense of humour and we later agreed that the most healing moments had been hugging giant redwoods and the superb sea views and hot chocolate at the Nepenthe restaurant in Big Sur. This was the first of many trips together that we managed to work into my filming schedule: an exhausting, nightmare commute between two completely disparate worlds.

Jenny needed to be closer to her doctors than out at Ulu, so I bought a house for us in Langata, one of Nairobi's leafier suburbs. Since my mother had died my father had been living in her old house, and a couple of years later fell ill with stomach cancer and moved in with us. Sitting out in the sun one day, he had watched helplessly as my vehicle, which I had foolishly entrusted to the Land Rover's infamously bad handbrake, rolled past within inches of his feet and down the hill to crash into a tree. That was one of the few laughs we had, and less than a year later he was bedridden and dying.

Sitting with him towards the end, as he moved in and out of consciousness, I remembered some of the few things I had shared with this stern, hard-working man. Because of the war, I had not seen much of my dad until I was about ten. I had no memory of bedtime stories, but I remembered the wonderful toys he had carved for me, my visit to his factory and that short safari with a sip of beer when I was twelve. But then, with a rush of affection, I suddenly recalled that when I was a little older he had introduced me to the Yukon gold-rush poems of Robert Service, especially 'The Cremation of Sam McGee', which he had recited to me from memory. Calling several neighbours, I managed to find a copy of Service's book, and though on revisiting the poem it seemed rather macabre in the situation, I knew he had loved it. Leaning close to him I started to read: 'There are strange things done in the midnight sun, by the men who moil for gold' – slowly the pain seemed to drain from his face and by the time I got to the end – 'The Northern Lights have seen queer sights, but the queerest they ever did see, was that night on the marge of Lake Lebarge I cremated Sam McGee' – he was smiling. That night he died quietly, leaving me with too few memories, deep gratitude for bringing me to Africa and a last peaceful smile.

I had chosen to set up the base for the Zaire project at Epulu, in the great Ituri forest. It had been on Stanley's route, where Lang and Chapin had collected, and where Johnston had discovered the first okapi, the strange forest giraffe that I hoped to film. I was among the ghosts of some of my heroes. Epulu was also the site of the old Belgian capture station, which had once supplied breeding groups of okapi to the world's best zoos, but had been looted and abandoned in the sixties. I had permission from the National Parks to build a little house on the banks of the Epulu river downstream from what was left of the station. Just two bedrooms, an eating area, an open-sided kitchen and a pit lavatory. I would bathe in the wide Epulu river, watched by giant kingfishers and collared pratincoles that hawked over the rapids. As I needed to be able to fly in and out, the Warden allowed me to make a very short airstrip along the forest edge. I would also need helpers, so took on Rogé, who had headed up the airstrip work, and a village girl whose only Western garment was a large pink petticoat, which she wore over her traditional wrap, and who claimed she could cook beans and rice and *sombé*, the spinach-like leaves of manioc. This proved to be only partly true, but I employed her because of her name – Vomis. I couldn't resist having a cook called vomit! I was greatly missing Kiarie, our old Kikuyu cook, who had known Joan since she was born and had stayed with her when we parted. Despite his great fear of wild animals Kiarie had accompanied Joan and me on all our safaris and had invariably produced hearty meals whatever the conditions. He had baked his crusty bread in an old ammo-tin oven, watched by huge, jet-black and naked Turkana warriors on Lake Rudolf, had shown some Maasai how a pressure cooker worked, amused by the way they bolted when it suddenly blew off steam. He had camped out among every kind of, for him, frightening animal; his eyes always wide with wonder at the way of life the little girl he had watched grow up had chosen to live. For many years we had been woken every day in camp by the low zzz of Kiarie slowly unzipping his tent, before getting up to rekindle the fire and make tea. Then one day there was that slow zzzzzzzz, followed by an explosive roar and then a zeeeeep! as he whipped that zip back down at high speed. Kiarie's unneeded explanation came small and nervous through the dark, 'Simba Bwana' – two KiSwahili words that I'm sure need no translation.

Shortly after I settled in at Epulu, a Swiss couple, Karl and Rosy Ruf, arrived to rehabilitate the station and resume capture operations. The okapi in captive breeding groups around the world were in need of fresh blood and a large American conservation organisation was funding the rebirth of this unique outfit. An energetic American couple, John and Thérèse Hart, had just started a research project on the okapi, and were working with the Mbuti to capture some in pits and radio-collar them. Filming the operation and being involved in the discovery of new information about the elusive okapi was an amazing opportunity. I would be lucky if I caught even a glimpse of a wild okapi, and getting film of their behaviour was almost certainly impossible, so although it would be a departure for me to make a film about people doing research, the combination of the Harts and their three young daughters, the Mbuti pygmies and the iconic okapi, was interesting enough to warrant a change of tactic. It would be the easiest subject on which to make a start, and a good introduction to the forest. The Harts were about to set off with porters and family to their research camp at Afarama, twenty kilometres into the forest, so I joined the caravan. Sarah, their serene twelve-year-old daughter, usually went way ahead on these journeys, to have the quiet forest to herself. Then came John, his enormous calves pushing him effortlessly along, followed by Thérèse with four-month-old La Jo in an improvised sling on her back. At the end of a line of porters came eight-year-old Bekah, usually with a couple of African friends for company, shrieking and giggling like a flock of parrots.

Their camp at Afarama was a collection of mud-built rooms furnished with narrow beds made of forest poles and roofed with iron sheets. They were spartan, but dry. This was the beginning of their research season, so the next day there would be a blessing ceremony known as Kutambikia, in which a large fire is lit below a tall tree and each man brings a bunch of leaves from the most important species in the forest, plunges them into the fire and washes himself in the smoke. It is a celebration of the forest and a way of seeking the approval of the spirits. Hands swirled sinuously in the fragrant column of smoke, shot through with shafts of sunlight, while Luwee, an old hunter, called out a litany of renowned ancestors and important plants, throwing on a large leaf for each name for the smoke to carry up to the spirits. Luwee told me that the forest welcomed me, but after all my years on the sunlit plains and open savannahs of East Africa, I felt an almost instinctive fear of the gloomy, humid rainforest, with its disorientating, foreshortened views, its mouldering smells, strange new sounds and the sheer size of the trees that towered above me, hiding all of the sky. It would be some time before I felt at ease in these new surroundings.

Over the following weeks two hundred narrow pits were dug, two metres deep and covered in a framework of slender sticks, on to which was laid a layer of the large *mangongo* leaves with which the Mbuti thatch their houses. With a final scattering of earth and leaf litter as camouflage the pits became virtually invisible. The Mbuti knew the exact location of every one, but we less aware palefaces fell into them with some regularity. On the Omo river in Ethiopia I had once half-fallen into a pit dug for hippos, with only the strength of the covering sticks, built for heavy animals, saving me. That pit was designed to kill and I had been hanging, kicking madly, above several long, pointed stakes. On the three occasions that I fell into an okapi pit that memory came rushing back as I lay on the soft earth bottom, relieved to find I had not been impaled. No one was ever hurt; and the Mbuti found it hilarious, but then the Mbuti found everything hilarious – for ever laughing or singing – I have never known a more happy-go-lucky people. The pits were checked twice a day and often contained a bush pig, various species of duiker antelope, mongoose or pangolin, which we would release. At long intervals we would find an okapi in one, too. These gentle creatures stood with their heads poking up above ground level, making no effort to escape and calmly gazing around as people crowded in to fit the radio collar and just to get a close look at these almost mythical

creatures. Its head is unmistakably giraffid, with huge liquid eyes, long muzzle and an even longer dark tongue that snakes out to groom its cheeks, ears and eyes. The males are slightly smaller than the females and have two short, hair-covered horns. Their elongated neck widens to a rather tubby, horse-like body and powerful hindquarters, which display the okapi's extraordinary markings. The animal's short, velvety pelt is uniform rich red brown but on its back end – like a sunburst radiating from its tail – it has a remarkable pattern of vivid, chalk-white stripes, plainly visible when it is moving away – which is just about all that is ever seen of an okapi in the wild.

After a couple of months I felt more at home in the forest, confident enough to take fairly long walks without fear of getting lost. I spent all my spare time soaking up the ambience of the aquarium light and cathedral space. Above the constant background buzzing of insects, the strange new bird and animal calls that I had yet to identify rang sharply in the warm, unmoving air. Compared with the savannah, birds and monkeys are difficult to spot in the forest, and are seen mostly as dark silhouettes against the sky. All that peering upwards puts a great strain on the back of the neck. But often I sat entranced just by the insect show. With late sunlight shafting through the trees like theatre spotlights against an almost black backdrop, in a patch of sun on the forest floor a gaggle of butterflies flashed wings of turquoise, yellow and carmine as they jostled for space to suck the minerals from the spot where some creature had peed. A large brown ant bit fiercely at the spine of a leaf, its brain programmed by a fungus that drove it up to precisely this spot where it would take a death grip with its jaws and die. Soon fungal threads would sprout from its body and shower their spores to infect any ants below, who would repeat the process. I watched his robotic behaviour and wondered about this new environment, where the most danger came from tiny creatures – flukes and filarians, plasmodiums, amoebae and nematodes. I was bound to pick up some of them, I thought. Fifty feet above me a goliath beetle buzzed – with that familiar motorbike sound that would have made my old mum smile – around a cascade of ripe figs, disturbing a bright chestnut squirrel. Then, in the luminous, mote-filled air, I watched a strange aerial display by a flock of large flies: after hovering stationary for a while, they climbed vertically for about twenty feet, then popped out their wings to form a surprisingly large, silvery parachute on which they descended very slowly to their original level. Up and down they went in a fluctuating, oscilloscope dance that I imagine impressed each other, or watching female flies, as much as it entranced me.

The Mbuti get their protein from the forest and eat almost anything they find or catch. Knowing my great interest was animals, they began offering me creatures that were on their way to the cooking pot. I made it quite plain that I would only pay them what the creature was worth as meat at the market. I did not want to encourage them to bring animals to me because it was more profitable than the hunting for the pot that they were doing anyway. It was a difficult line to tread, for they would hold up some endearing creature and mime eating it with great relish, rolling their eyes and smacking their lips, knowing I found it hard to say no. But I also had to be pragmatic, if a bushbaby or mongoose was on their menu that night, well, so be it, this was the way the Mbuti had always lived and as long as they hunted only for themselves, and not commercially for the passing truckers or miners, the forest could provide. However, young animals that I might rear and eventually be able to film I took in and soon we had a bustling menagerie. We built a large aviary around a tree that we had to pass through to get into the house – the human traffic helping the animals to get used to us. We had acquired a potto – a two-pound, moon-faced primate, a hunched-up ball of mahogany fur hiding deceptively long and powerful limbs – who moved slowly to the food tray, where it chose tree resin as its staple diet above all else. It soon enjoyed being scratched along the strange row of vertebrae covered in thin bare skin that projected between its shoulder blades. It would hunch its back to present these spikes, and tuck its head under in invitation, rocking back and forth and emitting a wonderful musky scent. Our tree pangolin – cousin to the one that my mother had agreed was a fish and covered in a thatch of scales like an elongated pine cone – gave out a similar odour. At two feet long, more than half its length is made up of its slender prehensile tail; this pangolin is an expert climber, tearing open the carton nests of arboreal termites and ants with powerful curving claws, and slurping up the insects with its long tongue. Ours had a baby that rode on the base of her tail as she clambered around the tree, and slept rolled up inside her curled body like the sleeping bag inside a bedroll. These were shy, nocturnal, tree-dwelling creatures that I knew I would probably never film in the wild. One of the Zaire films I planned to make would be about all these extraordinary rainforest creatures, many of which had never been filmed or some even seen by white men. I knew that it would take many years to film all the species I wanted, so I fitted them in wherever I could.

A lorry turned up one day with some adventurous young

tourists on it, one of whom, Karin Lagerstrom, was a veterinary nurse whom I persuaded to stay to help look after the growing population in our aviary. At the end of the project she did a wonderful job camping out deep in the forest with a whole range of creatures and successfully helping them go back to the wild. Filming captive animals can arouse great indignation among some purists, but the reality is that it is almost impossible to film sequences in the wild of many creatures that live deep in forests. The incredible advances made in digital photography have made filming in the darkest of forests possible today, but the limitations imposed by the shyness of the animals, fleetingly glimpsed among the dense vegetation, still apply. Some species, mostly primates, may be amenable to filming because they have been habituated – read tamed – in long-term studies by some patient scientist, and in some areas further west on the continent some species come out into clearings in the forest to feed and find minerals.

Apart from those specific animals, in the deep forest the best you can hope for, after a long, long wait, are some cursory scenes or bland shots from a camera trap. I wanted more than that and feel to this day that the unique behavioural sequences I was able to capture with creatures that would otherwise have been a pygmy's dinner, validated the way in which they were obtained. The point needs to be made that these animals were not trained: they were just tamed and were behaving naturally in a natural setting. At a film festival I was once taken to task by a man who said he had been filming for ten years and had never needed to use a tamed animal. After I had heard him out, I pointed out that he had filmed mainly elephants and lions in Kenya's national parks. Quite apart from the ease with which those animals can be found and filmed, he was benefitting from fifty years of visitors and researchers to those parks who, by their quiet attempts to get closer, had been taming them for him. If I had determined only to film those forest creatures in the wild I would certainly still be sitting in my tattered hide, frustrated and fungus-covered.

After a couple of weeks of heavy rain, many forest trees were now in blossom, a froth of white on the treetops, seen only from a hilltop or from the air. Flying low, with the doors taken off for filming, we were bathed in warm perfume, the morning air filled with the glint of bees. The honey season was coming, and the Mbuti camps were warming up with anticipatory song and dance, drummers beating a fierce rhythm as a shuffling line of dancers snaked around among their huts. Ancient crones and white-daubed children wagged their shoulders from side to side, holding adzes or knives, the honey hunter's tools. A young man mimed the search for honey: cocking his head as he listened for bees, craning skywards as he scanned the tree-tops for nests, and playing a honey pipe as a herald to good hunting.

The day after the dance I watched in disbelief as two young men, Bulaimo and Lomba, started to climb towards a large bee-filled hole a hundred feet up in a massive brachystegia tree. Its lowest branches were about seventy feet up, so they first had to climb a smaller tree whose top branches fell some ten feet below the lowest of the giant's. With their arms half round the trunk, legs folded frog-like and feet splayed out to the sides, their technique was as simple and effective as that of an inchworm caterpillar: reach upwards and grip, arch your back and bring up the rear, grip with the feet, straighten your legs and back, and extend upwards. Lean as skinned rabbits, their backs flexed and straightened, the well-defined muscles sliding under the skin, they went up that smooth trunk almost as easily as you or I climb stairs. One carried a long coil of liana, a flexible forest vine, and a small woven basket, the other a bundle of green leaves secured round some glowing coals, and they moved rapidly through the branches to emerge some ten feet below the honey tree's first heavy limb.

Using a long thin branch, they then manoeuvred the liana over the fork, secured it and clambered up it into the canopy, where the bees were already disturbed by their approach and were buzzing angrily around their hole. The men were getting stung but Lomba quickly blew on his coals and clouds of white smoke soon disorientated the bees, which swirled around them but no longer stung. Lomba kept the smoker secured round his neck, knowing that if he dropped it the bees could attack en masse and cause him to fall to his death. Pushing his arm into the hole, Bulaimo removed dinner-plate-sized pieces of comb, heavy and dripping with honey. He filled his basket and lowered it to the ground with a liana. These first combs of the season go to the hunters, so the two young men came sliding down the tree at speed, leaving a column of bark dust floating in the light to mark their passage, hurrying to get their share.

Replete and relaxed Bulaimo and Lomba asked if I would like to join them and a group of Mbuti who were about to leave for a camp in the forest to do some net hunting. Collecting more film from the house, we set off and after a five-hour walk arrived at the camp, a random collection of small igloo-shaped huts, some still under construction, with smoking fires, children playing and adults engaged in loud conversation and various tasks.

A woman stood inside a wide framework of slender branches, planted in a circle then bent inwards and tied together at the top and encircled with horizontal saplings. From a pile of large *mangongo* leaves she took one at a time and fixed them like tiles on to the lattice, folding each stem and pushing it back through the leaf. She worked quickly, giving the leaves plenty of overlap – this house would stay dry. Another woman was singeing the hair off a hyrax before cooking it, while another mixed ash and the milky sap of a liana to paint her small boy's face with chalk-white designs while an ancient lady, deep in concentration, painted lines radiating out from her flaccid dugs, turning them into vivid starbursts. Several men sat chatting quietly and twisting slender strips of dried liana into string by rolling it on their thighs, which, when I tried to do it, I discovered is a painfully depilating operation for a hairy-legged white man. They weave the cord into nets about the height of a tennis net, but larger meshed and up to eighty yards long. They were checking and repairing the nets, rolling them into skeins to keep in a dry place ready for tomorrow. The following day we were woken by the rumbling croaks of colobus monkeys and set out early, with nets looped over the hunters' heads and shoulders, to assemble at a central point where the hunters split into two wings.

In the deep forest, with visibility limited to a few yards, the organisation of a net hunt is almost telepathic, and we palefaces would certainly need GPS units and radios to position and coordinate ourselves in a huge semicircle as precisely as do the Mbuti. The left-hand wing of nets is known as *mangé*, the bow, which is held in the left hand. The nets to the right are named for the weapon the hunters hold in their right hand, *ekongaa*, the spear. Each net bearer moves swiftly, uncoiling his net on to the ground in a long, slightly curving line before returning to hold it down with his foot, lifting it to the right height and securing it to twigs with a dexterous twist. On each wing three or more nets are set in a wide arc, while the women, older children and men position themselves some distance away, facing the mouth of the semicircular trap. Tense and expectant, I waited with a young hunter, Kayo, at his net near the middle of the line. He had set it well, now success would depend on the charms that hung from it – the shining seeds and guinea fowl feathers – and on the beaters. The forest was silent but for the honking and whooshing wingbeats of a passing flock of hornbills, then a wave of sound came drifting through the trees. Whooping, shouting and beating the vegetation, the line of beaters moved forward, scattering everything in their path, with bursts of wild hooting rising whenever an animal was spotted. The noise got louder, closer, '*Mboloko, mboloko!*' (blue duiker) and suddenly that small animal streaked into view and punched into the net, forming a deep bag that firmly entangled it. The hunters were up and running along their nets, spearing a large bay duiker, a further two blue duiker and a pygmy antelope, a dainty, foot-high, five-pound animal that is the smallest ungulate in the forest. A net close to the far end caught a water chevrotain, a short, stocky semi-aquatic antelope with sharp canine tusks for which the Mbuti have great respect. It is a creature I long to film. The Mbuti tell me it rests in hollow trees and eats fallen fruit, flowers, mushrooms 'and

everything'. They also say it is so shy that I will never film one. They did promise, though, that if they caught a young one they would bring it to me alive, 'as long as we have plenty of other food that day!'

On the way out of the forest I stopped to listen to a distant roar that sounded like an approaching jet, and finally realised that the noise was wind and rain hammering on the canopy as a storm approached. The Mbuti told me that the forest is a dangerous place to be during a storm, as old branches are liable to fall. They broke into a run and we raced the last mile to the forest edge and down the road to Epulu.

Those storms were not only a frightening hazard when in the forest but even more so when flying. In East Africa it is usually possible to fly around storms without deviating more than a few degrees, but these Congo storms were often seventy or more miles across. Deviating that far puts you many miles off course. Fifty miles east of Epulu, on the edge of the Ituri forest, was Nyankunde, the operations base for the Missionary Aviation Fellowship, who provided air support for the many missions scattered across the country. This was a village transplanted straight out of middle America, with children on Schwinn bicycles, motorised lawnmowers, 110-volt refrigerators and ice-cream machines, women in gingham, apple pie and a fine hospital. More important, there were aviation fuels, skilled Cessna engineers and pilots who knew the country. On my first visit there, well aware of the problems I would be facing, they had photocopied their 'bible' for me – the details of all the small airfields they used, but also the position and compass headings of all the bridges, which they then explained by pointing out that, unlike the roads, bridges cannot get hidden by trees. Many were broken or unused, but they were still easily visible and, unlike roads and rivers, they didn't meander and all of them faced a slightly different direction. By flying low along the bridge and getting its compass bearing you could identify it and thereby get an exact fix on your position. It wasn't much, but in those days before GPS, it was something.

Towards the end of my time in Zaire, after eight years of hard navigation, I bought a newfangled Trimble GPS, to which one of the MAF pilots kindly transferred every airfield and bridge. Just weeks later, on only my second trip with this new backup, I had been seventy miles from my destination when hot oil spurted down my leg. The pipe to the oil pressure gauge had broken and my engine oil was being pumped away. Long before I got to Epulu it would run out, the engine would seize and I would go down. Swallowing hard, I punched 'nearest' into that little gadget and magically it gave me the heading for a tiny, seldom used strip I had been unaware of just thirty miles away off to my right. By the time I landed every drop of my oil had been transferred to my legs and shoes, and the missionary who came out to meet me wondered why I was kissing a small grey box.

Before that box, I had made several terrifying flights in Zaire when I was pushed many miles off course by storms, my little Cessna 180 flung around like a dry leaf, rain drumming loudly on the wings and dripping on to my map, as I ran low on fuel, daylight and experience – protractor, ruler and pencil taking turns to slide to the floor as I struggled to work out my position by dead-reckoning, my frightened eyes going back and forth between the gaps in the rain, the map and the fuel gauge. If I went down I would disappear into that big broccoli like a pea tossed on to a lawn, and even if I somehow managed to put it down on the canopy and the plane was held there, I would not be able to climb down the tall, branchless trunks. I'd have to starve in the treetops or jump sixty feet down to the army ants. Nor was there any point in flying high, where I might be able to glide to a safe landing place, or make a 'Mayday' emergency call: there were no clearings and there was no one to answer the radio. Being high meant only that I would have more time to scream on the way down, so I opted to fly low, where there would only be time for those words that I believe are common to almost all black box crash recorders – 'Oh shit!' And flying low over the forest was such an exhilarating joy. Flocks of hornbills burst from fig trees, scattering their black-and-white patterns across the green canopy like dice flung across the baize, there were scarlet splashes that were new leaves or the tails of grey parrots, the sight and scent of mile upon mile of blossoms and, after rain, bedraggled groups of colobus or red-tailed monkeys on the top of emergent trees, warming up in the sun.

The fierce winds that came ahead of storms were always frightening, sometimes terrifying: violent, wing-wrenching gusts that would fling my plane up into the leaden cloud base, or down to the very treetops. Conventional pilot wisdom has it that downdrafts form a cushion when they hit bottom, so a plane is unlikely to be driven right down to earth. But to the north of Epulu there were two areas that belied that story and spoke strongly of the immense power of the winds. Along a path of devastation maybe a hundred yards wide and eight hundred long, every tree lay flat on the ground, all pointing in the same direction. Except that they were green, not burned black, these areas looked exactly like the flattened forest on the slopes of Mount St Helens, with hundreds of giant trees lying prone and parallel on the ground. Nothing else could have caused this but wind and the phenomenon – linked to severe storms – known as a microburst. Dry air, dragged down from thirty thousand feet by heavy rain, accelerates downward to create a focused burst of wind that slams into the ground at up to a hundred and seventy miles an hour, flattening all in its path. Hit by one of those while flying low there wouldn't even be time to sa . . .

True to their word, the Mbuti one day brought in an *afélé* – a young water chevrotain – that had been found in a hollow tree, a sturdy antelope built like a goat, its coat a russet brown with large white spots on its back and a wavy double white line along its flanks. She took to the bottle readily, soon became very tame and, along with several other creatures, had the run of our forest house. It would be a year or so before she matured and I could film her, but she was going to surprise me.

Constantly underfoot in our house was a rat-sized, hunch-backed creature with kangaroo-style hind legs, a rat's tail and a long flexible nose with which it endlessly rummaged in the leaf litter I had piled in the corners of our rooms. Its proper name was the chequered elephant shrew, the Mbuti called it *akbeké*, we called it hose-nose. Chestnut, with rows of white patches down its back and large, bright eyes, it wielded its highly mobile nose to flip leaves and sticks aside in its search for insects. Pushing its nose under a heavy stick, it would thrust up forcefully with its powerful

hind legs, hurling the stick a foot or more into the air. Unfortunately it was usually straight up into the air, so the stick would fall back across its long proboscis, leaving hose-nose with a puzzled expression and the urge to try again.

The shrew I most wanted was an animal I had known about since boyhood when I'd read about Gerald Durrell and his Bafut beagles trying to catch a giant otter shrew. One of Durrell's crew had been bitten and his yell, 'Ayee! Dis blurry beef done chop me!' was for years the way my young friends and I greeted similar accidents. Otter shrews spend all day in a burrow and most of their nights in the water, and after many long nights trying to catch more than a glimpse of one I decided that trapping was my best bet. The Mbuti had brought me the drowned body of one of these two-foot-long animals he'd caught in a fish trap, and said this was the best way to catch them – but catch them dead, which was not what I wanted. So after some thought I imported an eel trap, in which two wings guided the migrating eels into a ten-foot-long tapered, tubular net, held open by metal rings, punctuated with a series of funnels that guaranteed one-way traffic. I took the end of the tube up on to the bank and fixed it round the entrance to a live-catching raccoon trap, baited with fish. It was an effort to set up and we had to try it on several different streams, but within a couple of weeks we had successfully caught a live otter shrew.

I was concerned that Durrell's animals had died after only a short time in captivity, their coats wet and matted, so I designed a sleeping box based on one I had seen in Australia for a platypus – a creature that also gets sick and dies if unable to dry its coat. I fixed a piece of inner tube across the entrance hole with a narrow, horizontal slit, so that when the animal forced its way through all the water was thoroughly squeezed from its coat. Dry and warm between swims, our otter shrew thrived. With a sleek brown pelt, long flexible tail, broad head and large whiskers it hunts crabs and fish below water, coming to a sandbank to eat, which it does clumsily, apparently unable to hold down its prey with its unwebbed front feet. Our *akbedu* lived with us for several months and after filming him we released him back into his stream. He didn't get any tamer in all that time and often went for us with his impressive teeth. We had no accidents, but I know that had anyone been bitten we were all ready to yell, 'Ayee! Dis blurry beef done chop me!'

10

Going Home

Every two or three weeks I would fly to Nairobi to spend some time with Jenny and to resupply our base at Epulu, and on one of those trips I learned that Bernhard Grzimek had just died. He had taken a group of children to the circus and had simply fallen asleep in his chair. Over the years, using the platform of his TV programmes – the most popular and influential in Germany – he had raised over a hundred and fifty million dollars for various conservation causes, had effectively tackled ministers and presidents on conservation issues, and fought fiercely against the bludgeoning of baby seals, wearing spotted cat furs and the killing of rhino and elephant. I had worked with him for almost thirty years and was proud that he acknowledged the major contribution my films had made in his fund-raising campaigns.

His ashes were buried alongside his son Michael on the rim of the Ngorongoro crater, and I was honoured to be asked by his family to deliver the eulogy. This time, instead of the five or six colleagues who had buried Michael, there were dozens of the great and the good from around the world, come to pay their last respects to one of the most visionary conservationists of our age. With fantastic views of wild Africa from the graveside, I spoke to the crowd: 'The inspirational architect Sir Christopher Wren was buried in St Paul's, the sublime cathedral which he designed. His headstone reads: "If you seek his epitaph, look around you." So, look around you, Ngorongoro and the Serengeti, wonderful places that today are better known, and better protected, thanks to Bernhard Grzimek. These are his epitaph.'

He was an extraordinary man, and I am fortunate to have known him so well and to be able to call him a friend. I later heard that, in one of those inexplicable events that raise the hairs on the back of your neck, the three largest-maned lions in the crater climbed up to the rim road, something they very seldom do, and lay for many hours in front of the grave.

On many of those trips back to Nairobi I also had to call in to check up on the Serengeti project. Mark Deeble and Vicky Stone had now joined the project and were soldiering on with very little guidance from me. We had dropped the idea of a series, and instead they were concentrating on the huge crocodiles in the Grumeti river and the annual mayhem caused when the wildebeest migration arrives. I took some time off from my project in Zaire to join them filming there and later as they were filming along the Nile. Their finished film, *Here Be Dragons*, was a great success that I'm happy to say set them on the road to becoming the best in the business.

Wildlife filming is captive to the seasons and it was now essential that we move from the forests on to the next location. We headed to the far north of Zaire, up against the Sudan border, to a national park unique in Africa – Garamba. We welcomed the move to open savannah country, but unlike the short grass and tall trees of the Serengeti, Garamba was in a vegetation zone that ran right across to west Africa, known as Guinea Savannah, with ten-foot-high grass and trees that seldom exceeded thirty. That grass was now tinder dry and we had to be ready for the huge fires that sweep across the park every year, driving elephants and every other species ahead of the wall of flames. When the Garamba Warden heard I would be flying up he asked if I could help by bringing the wages for the park rangers, which were long overdue. In my ten years in

Zaire the local currency, the Zaire, went from sixty to the dollar to thirty-six million. I found it impossible to keep tabs on my budgets, but somehow the village women in Epulu worked out the new price of their produce each week, though they were happier bartering avocados for aspirins, or bananas for biros, while the banks handed out briefcase-sized blocks of decomposing notes that no one could possibly count. Four months' wages for twenty men completely filled my six-seater aircraft. There is a well-known tale – possibly apocryphal – that on seeing Mozart's opera *Il Seraglio*, Emperor Joseph's comment was 'Too many notes'. He really should have seen this lot.

In 1879 King Leopold of Belgium ordered that a project be set up in the Congo to capture and train African elephants, something that the rest of the world said was impossible – but history told otherwise. Around 300 BC the Romans had begun plundering the upper Nile area for animals to use in their coliseum games and had started shipping African elephants from Meroe, in the kingdom of Cush (modern-day Sudan). From there the animals were walked to the Red Sea, then transported on purpose-built, shallow draft ships, via several dedicated ports, and marched on to Rome – twelve hundred miles, a truly elephantine undertaking. The best-known military campaign using African elephants was in 218 BC when the Carthaginian general Hannibal marched on Rome with forty elephants that he took through a ten-thousand-foot pass over the Alps, and on rafts across the Rhône. The Romans used them rather like tanks, and they were involved in many subsequent battles and in much greater numbers than with Hannibal, but then, for some reason, that culture died out, and some two thousand years later all that experience with African elephants was forgotten and only the Indian species were considered trainable.

King Leopold's plan involved shipping four Indian elephants from Bombay to the coast of Africa, in what is now Tanzania, from where they would be walked to the Congo to start training their African cousins. With slavers raiding over much of their route, and human and animal disease rife, this was not a good time for a technology transfer of this sort and within a year, still short of the Congo by hundreds of miles, the last surviving Indian elephant died, so the mahouts said, 'of sadness'. Leopold's scheme was delayed for twenty years, but when it was revisited the new plan was to capture African elephants and bring in experienced Indian mahouts to train them. The operation was run on strict military lines by a Belgian cavalry officer, a certain Commandant Laplume, and by the 1930s the station at Gangala na Bodio, in Garamba, had one hundred and twenty fully trained elephants, pulling wagons and ploughs at one tenth of the cost of tractors. I had managed to find some contemporary black-and-white film shot by my mentor, the late Armand Denis. Muscular Azande men raced alongside a herd, roping a half-grown animal and finally subduing it, then to be led away by two older 'monitor' elephants. In the film, huge covered wagons were pulled across rivers crammed with hippos by teams of elephants and a cavalry officer on a white horse inspected sixty elephants parading in immaculate rows. The film was a window into an incredibly romantic and adventurous time, for ever gone, and I pressed my nose against it longingly.

I was now headed to Garamba to film the history of the park since those first golden days. Much of the story is grim. In the fifties there were around fifty thousand elephants and perhaps two thousand northern white rhino in the park. By the seventies ninety per cent of those rhinos had been killed by poachers and the elephant population was reduced to twenty thousand, and when two good friends of mine, Kes Hillman and Fraser Smith, arrived at the park in 1984 there were a mere four thousand five hundred elephants and just fifteen rhino left. Kes and Fraser are an incredible couple, unsung heroes in one of Africa's fiercest conservation battles. Kes is an English biologist who came to Africa with a truckload of twenty-something tourists in 1979.

A trim, pocket-sized Venus, usually dressed in old military gear and boots, she looks like a poster girl for the Israeli army. Fraser had been a game warden in South Africa and after they met he moved up to the Congo to join her in Garamba. By the time I arrived they had been more or less running the park for a decade, Kes doing research and monitoring the white rhino population, while Fraser handled the technical side of conservation together with Muhindu Mesi, the Zairean Warden. Muhindu was known to be a tough customer, but at our first meeting I thought he was carrying tough a bit far. His Garamba office was in an old colonial building with the very high ceilings the Belgians had favoured. The whole ceiling was now a moving, twittering carpet of bats that shuffled excitedly as I came in. The floor was liber-ally splattered with fresh guano and the air was like breathing ammonia. In the middle of the room Muhindu sat stoically at his desk, sheltering from the storm under a huge Cinzano umbrella. With watering eyes and wheezing breath I greeted

him, standing as close as I could to the shelter. But because of Muhindu's big desk I could only get to the edge of the umbrella, where I collected all the run-off. I was cleaning my glasses for the second time when Muhindu suggested we talked outside.

The guy had style, and was an honest, hard-working man who cared about his park and its animals. If caring were all that was required, the devotion and hard work put in by this trio over two decades would have made Garamba a huge success. I was fortunate to be there when it was making a comeback from desperate times, but it wasn't easy. Fraser was putting in roads, bridges, communication networks and ranger posts; Muhindu had trained a competent ranger force and Kes was monitoring the slow but steady increase in the white rhino population. There were still four old trained elephants alive, all in their fifties, and a couple of aged mahouts who remembered how to handle them. With no reason for them to be ridden, the elephants had spent those years hanging around the park headquarters, sometimes chained by the leg, sometimes free to wander. Among all their other projects, Kes and Fraser embarked on a mission to retrain the old elephants for riding, and to capture and train some young ones.

Garamba is a spectacularly beautiful area of wide grasslands with scattered trees, gallery forest, many springs and rivers swarming with hippo. With elephant and buffalo in herds of hundreds, antelopes, lions, white rhinos, Zaire's only giraffes and some spectacular bird life, this should have been a prime tourist destination, but with no connection by air, and a week of driving on terrible roads away from the standard East African tourist circuit, Garamba was just too remote to attract any visitors. But, went Kes's and Fraser's thinking, if Garamba could provide the unique experience of riding through that wild country on elephant-back, perhaps there was a chance of attracting high-end tourism and revenue for the park. We all hoped that my film might start that ball rolling.

My time there was all adventure, ranging from tracking rhinos on foot and shooting them with a special dart that cut a tiny plug of skin from the animals so that they could be DNA-tested, to some hairy anti-poaching operations and the capture of young elephants for training. But the most rewarding moment of all was when we took the four aged trained elephants out into the park, with riders on their backs for the first time in decades, to show what an elephant-back safari could be like. Fraser's father and a group of friends were visiting at the time and we persuaded them to be guinea pigs for this potentially dangerous operation. We were not sure how the trained animals would behave carrying a howdah and six passengers, nor how the wild elephants would react when presented with such a sight: but it was a once-in-a-lifetime opportunity and everyone agreed give it a go.

I was filming from the back of one of the elephants as we set off, the nervous but excited passengers hanging on tightly. Our elephants crossed rivers belly-deep, pushing through large schools of hippos that snorted and cavorted around us. They took us to within yards of groups of unsuspecting white rhinos, past wary and puzzled lions, and – best of all – through great herds of their own wild relatives. Those wild elephants accepted

ours walking alongside some forty yards away, but I could see they were nervous, unsettled by the strange growths sprouting from our animals' backs. One group of over a hundred got our wind and, no doubt with recent memories of encounters with poachers, they raced past our animals to join the rest of their herd, thundering by in clouds of dust, trumpeting and making threatening gestures at us as they passed: unexpected footage that made a great film finale.

Considering that these old elephants had been mothballed for thirty years, they put on a remarkably calm performance, to the enormous relief of their first-time riders. The first two young elephants to be trained were also doing well, and were ready to show their paces when the Director of the Zaire National Parks came to visit along with a troop of politicians and some of Garamba's donors. Functions like this are what rank and position are all about for the Zairean leadership. They come with munificent travel allowances, local girls provided for the VIPs' entertainment and quantities of imported food and drink – pâté, brie and champagne – to accompany the heaps of meat, which was usually an animal shot in the park. And all of this was served by hostesses – city girls flown in for the occasion – who rather suggestively opened your beer with their teeth. The Director was so impressed by the newly trained elephants that he decided they should be sent as a gift to President Mobutu. When a suggestion like this is made in Zaire – if you want to stay in the country – you can only swallow hard and enthusiastically agree. So it was that a military transport plane arrived and two years of hard work, a steep learning curve and a careful training programme went down the drain. In Mobutu's badly run zoo, our elephants sickened and died. *C'est l'Afrique.*

Garamba's location, close to the Ugandan and Sudan borders, made the business of protecting elephant and rhino a nightmare. Various rebel groups operated close to the border, including

SPLR rebels in the Sudan, and Lord's Resistance Army rebels in Uganda, and they crossed into the park to hunt. Whenever there was a ceasefire in the endless civil war in Darfur, in Sudan, the terrifying Janjaweed horsemen would call a break in their slaughter of people and raid Garamba instead, riding west to the Central African Republic, to plunder ivory and rhino horn. These were well-armed groups that lived off the land, butchering helpless villagers for the hell of it and killing every animal they came upon. Fraser and Muhindu spent many dangerous hours trying to negotiate with these ruthless killers to stay away from the park. But the odds were stacked against them and things started to unravel when, in 1996, Rwandan-backed rebels advanced towards Kinshasa to oust President Mobutu.

One day a brand-new Pilatus, a million-dollar executive aircraft, landed at Garamba, closely followed by a huge, unmarked military transport aircraft that disgorged several tons of supplies, two jeeps, a freezer, a washing machine, three men and a dog. It turned out that Mobutu had recycled a couple of mercenaries who had fought for him in the sixties, and as a down payment in exchange for promising to help defeat the Rwandans they had been given Garamba as a hunting ground and retirement home. Fortunately Kes and Muhindu were away. Fraser was forced to leave his house and told that if he wanted to look after the park he would have to do it from out there somewhere. This he did, sneaking back once to rescue some personal effects from the house, his two cats and his private vehicle, which he hid in the bush. Then, watching in agony from the long grass, he saw the magnificent place that he, Kes and Muhindu had protected for so long start to die. Within weeks the Pilatus had been crashed and the old mercenaries, long past their best-before date, fled from the well-armed and experienced rebels, leaving behind tons of tinned food, arms and ammunition, and nearly two hundred drums of aircraft fuel, which the locals – with greater need of an empty drum than forty gallons of fuel – had simply emptied on to the ground. When Fraser was at last able to return, most of the park's buildings, including his home, had been looted and burned to the ground. Before long, the last of the white rhino were gunned down and Garamba, with such a long and fabled history, had turned into another of Africa's great conservation tragedies.

For ten years Jenny had been hanging on to life, going through a roller-coaster regime of remission and relapse, and I had been flying back and forth to be with her as much as I could in between filming. After so long and so much pain the possibility that Joan and I might get back together again had pretty well vanished and was not something we discussed any more. Jenny was not searching for an alternative cure now, more for a way to calm her mind and accept what was happening. Over the years I had joined her on all her pilgrimages. I have sat in the waiting room of the renowned healer Mathew Manning, looking out at his latest Jaguar, which seemed to be renewed at every visit. I didn't speak all day, chewed each mouthful of organic food thirty times, lay on the floor and chanted 'ohm' for half the night with Ram Dass and his disciples in England. I had the deeply humbling experience of kneeling before Mother Meera in Germany, having her touch my head, look into my eyes and attempt to fill me with light. I watched as a healer in the Philippines pressed his fingers into Jenny's swollen stomach and produced blood. I watched the total measure of her blood, pallid and sludgy with white cells, coming out of her right arm, going through a series of glass tubes and filters and back into her left – round and round in a hugely dubious and expensive process called polyatomic apherasis. (I always distrusted both the machine and the character who invented and promoted it, and years later I found it rewarding to learn that he had been jailed in America as a 'career criminal' who had 'left a trail of human misery in his wake'.) For their alleged healing powers,

I sought and bought sheep sorrel, essiac, burdock root, turkey rhubarb, and various 'rainforest treasures' and 'ancient seabed minerals'. Sadly, we never got to try dolphin-assisted therapy, in which I might have had more faith. Yes, I found it hard at times – I am a cynical old sod and can smell chicanery a mile off – but Jenny believed and my role was to support her. Thanks to her unwavering belief, many of these sessions did work, renewing her strength of body and mind, and actually, after two weeks with Ram Dass and his message of living in the moment, friends said I had not looked so at peace for years.

Returning to my base at Epulu after the Ram Dass pilgrimage, however, quickly wiped the serene smile off my face. There were riots in major cities and a mob rolled into Epulu looking for trouble. I was in the Land Rover with Karin, the Swedish vet who looked after our animals, when we found the road blocked by a shouting rabble who climbed on to the bonnet, stamped their feet and started violently rocking the vehicle. It felt as though they planned to overturn it, so I revved the engine, spun the wheels and scattered the crowd, fortunately causing no injuries. In Zaire it is sensible to expect and plan for the worst: I decided we should get out quickly. Back at the house we collected cameras and exposed film, a few belongings and our most dependent animals – two tree hyrax, the potto and a pair of hornbills – gave the staff two months' salary in advance, just in case, and flew back to Nairobi. We were able to return within a month, to find our staff and animals unharmed, but in ten years we had three evacuations of this sort – it was disruptive, debilitating and expensive.

Back in the rainforest, I was now chasing the remaining species that I was determined to find and film. Surprisingly, the okapi had been relatively straightforward thanks to Karl and Rosy Ruf's capture operations going so well, and there were now several animals in large enclosures in the forest. The Harts' radio-collared animals had led to an understanding of the okapi's movements and territory size, but none of us was ever able to observe them in the wild and all our behavioural information came from the captive okapi.

The water chevrotain also had the run of a large patch of forest with a stream running through it, but on the day that I had first filmed her diving into the stream she completely disappeared and I thought we had lost her. Waist-deep and worried sick, I was desperately searching under the overhanging vegetation when I felt her little hooves lightly walking across my bare feet. I was elated, not just because she was still with us, but having filmed hippos walking weightlessly underwater I now understood from those soft footfalls that that was exactly what she was doing.

Chevrotains were known to take refuge in water, where they hid under stream-side vegetation, but we had never suspected they might be able to walk on the river bed. With an underwater camera I was able to film that extraordinary action. With her head held low to avoid being lifted by the current, and her eyes open, she walked lightly along the bottom, a stream of shining bubbles escaping from her coat, scattering a school of fish. Stopping to lift her head to take a breath, she lowered it again, sank down and carried on walking. She was completely at home underwater and, like the hippo, she can only have stayed down like that by compressing her ribcage and reducing her specific gravity. It was a wonderful discovery and sequence, and before long I was happy to see her go off to freedom and her secretive life in the forest.

Zaire was wearing me down. Jenny needed more of my time and the overseas trips we were making in search of cure or comfort for her were emotionally exhausting. I was also being frayed in other ways, for although I hadn't been bitten by anything of note for many years, I had instead been stung by countless tiny creatures. I had played host many times to mango worms, the larvae of a fly with the wonderful Latin name *anthropophaga* – the eater of men. The fly lays its eggs on clothes hanging out to dry, and when the larvae hatch they burrow under the skin and start eating. They take at least a week, eating your flesh both to nurture and make room for their growing, maggoty bodies, before they are ready to crawl out of the boil-like lesions that have formed. You can actually feel them chewing at your flesh, and the itching and torment are awful. I was told they could be got rid of by blocking off their air with Vaseline, but it never worked for me. I had been down with malaria every year and was finally getting over river blindness – caused by little blackflies, with a burning bite, that were plentiful on the rapids in front of my house. Their bite is bad enough, but their saliva transmits microscopic filarial worms that soon multiply until millions spread throughout the body, reaching, in advanced cases, the eyes. When warmed up by exercise or a hot shower, these worms cause the most excruciating itching. I would take cold showers and lie on the bed to air-dry, desperately trying not to scratch. Many old Congo colonials would go back to Europe in the winter not for the mulled wine or the skiing, but to stand outside in a freezing wind and get respite from the tormenting itch of their 'filaires'. When these microscopic worms die, the body's immune system triggers inflammation and, when this happens in the eyes, it can bring blindness. Some thirty-five million people are currently infected across central and western Africa, with hundreds of thousands already blind. The pharmaceutical company Merck developed a drug called Ivermectin, which fixed the worms, but back then it was only cleared for veterinary use. However, Billy Karesh, the vet who came out regularly to check on the captive okapi, assured me it was safe for humans and gave me a large tube – meant for horses. I kept it in the fridge and every week squeezed out and swallowed a few centimetres, until slowly my itching had subsided. It has now been approved for human use and Merck has delivered millions of free doses in Africa.

*

It was 1993 and I was really ready to go home. For eight years now I had been asking the Mbuti – indeed anyone – for information about the two creatures I still wanted to film, the Congo peacock and the aquatic genet, but had had no luck with them. I was going to have to find them myself. I had a good idea of how to go about it, but I was tired and knew I would need help. I put out the word in Nairobi that I was looking for someone who wanted adventure and challenge, and did not need comfort, company or cooking. A friend recommended Giles Thornton, a young English adventurer who I was assured had boundless energy, and was very low maintenance. Giles was a lovely guy, with great style, able to make friends with anyone, and we got on well from the start. He knew little about African wildlife, but had tremendous enthusiasm and energy and the ability to focus it on any task. Working with him, it was if I had plugged jump leads into his great reserves of energy and I felt we stood a good chance of finding those creatures that had eluded Chapin, Lang and so many others since.

The only peacocks to be captured alive had been caught by Mbuti working with the animal trapper Charles Cordier in 1937, in an area a hundred miles west of my house at Epulu, so that was where we would start looking. I decided I would set up a base for the search at Opiengé, south of the potholed muddy track – punctuated with broken bridges and failed ferries – known as the Great Trans-African Highway. Although I had an incredibly competent six-wheel-drive Pinzgaur vehicle, fuel was scarce and on those terrible roads it was actually often easier to get around on a motorbike. Sometimes container trucks sank so deep into mud holes that you could step from roadside on to the container's roof. Two wheels also gave access to the forest trails that we used a lot, so it was on trail bikes that Giles and I set off to recce the route.

I had a dog-eared copy of *A Tourist Guide to the Belgian Congo*, a thick book that had been our bible when Joan and I had led photographic safaris through the Congo in the fifties and sixties. Back then we could plan our trips with complete confidence in the information that book contained. Turning those once-familiar pages was now cause for both laughter and deep sadness at this country's descent into chaos. Giles, who knew only the 'modern' Zaire, couldn't believe my stories of the old Congo and roared with laughter as we traced our route. We passed through NiaNia, where our guidebook told us we could send a message home via the radiotelegraph station, fill our tanks and have breakfast at the Hotel Mubali, which would cost us twenty-five or forty francs (continental or full English). Moving on along what had been the smooth earth road to Bafwasende, the guide promised us that Hotel de la Lindi offered a seventy-franc lunch and the chance to send another telegram. Next up would be Bafwabalinga and the turn-off to our destination, Opiengé.

'Why don't I drop you at the hotel and you can order the beers while I fill up and check the garage to find out what the Opiengé road is like.'

'OK, but you realise a room is two hundred francs and there's only cold running water? We should have stayed at Bafwasende!'

The jokes soon wore thin, however, for after less than forty years all that remained of the facilities detailed in the guide was rubble. 'What did the Congo have before they discovered candles?' the joke goes. Answer: 'Electricity!' The only restaurant we found among the ramshackle huts and shops at Bafwabalinga was an unsavoury truck stop where half an oil drum contained a grisly, gristly bush-meat stew of antelope and monkey chunks. Giles had a smoke while we watched a muscle-bound truck driver searching the stew for tasty pieces with a large wooden ladle. He first brought up a monkey's foot, then a grinning skull, leading to us dub the mix the 'walkie-talkie' stew after the South African poor man's meal of chicken feet and heads. We named the establishment 'MonkDonalds' and over the next month we saw quite a bit of that oil drum, which

sat over an eternal flame that kept it bubbling as new victims were thrown in. Eating our relatively safe bananas, we wondered what pathogens lurked in that primal soup, waiting to join AIDS and Ebola in the leap from simian to human host. Then one day we saw a fresh monkey paw emerge beseechingly out of the brew, an image uncannily reminiscent of that iconic shot of a soldier's hand reaching out of a trench in *All Quiet on the Western Front*. That was my last visit to MonkDonalds.

We had only come about seventy miles from Epulu, but already my backside was feeling tender from the ride and I marvelled at the baboons'-bum grade of calluses that the renowned mountaineer Bill Tillman must have developed on *his* rear end. For after climbing all the major East African peaks he had begun his trip back to England by riding the three thousand miles to Cameroon – on the west coast – on a bicycle. He had passed through Bafwabalinga in 1934 and as I climbed stiffly on to my well-sprung trail bike I thought, they don't make 'em like they used to.

We quickly discovered that Opiengé, an important gold-mining centre during the years of World War Two, when the Congo provided much of the gold that financed the Allies' war chest, could now only be reached by a footpath following the route that had once hummed with crucial traffic. I took my sore backside back to Epulu while Giles bashed on to Opiengé, where he would clear the old airstrip, allowing me to fly in several days afterwards with my bike, tents, rations and camera gear. Three days later my arrival was noisily greeted by *le tout Opiengé*, and a grinning Giles came out to greet me and to introduce Père Léon Mondri, the only other white face in the excited crowd. A missionary of the Sacred Heart of Jesus, Père Léon had lived in this tiny village for over forty years and hadn't ventured out for a decade. Short and stocky, with the powerful legs and shoulders of a scrum half and twinkling eyes under a shock of white hair, his energy belied his claim to be seventy-six and he laughingly admitted that forty years in Opiengé without a calendar could account for some leeway.

Within minutes of my tumultuous arrival, however, the crowd suddenly fell silent at the arrival of a menacing figure who claimed to be the local agent of a frightening organisation with an acronym worthy of a James Bond adversary: SNIP, otherwise known as the Service National d'Intelligence et de Protection. A large fellow in mirrored shades, this charmer led me to a dingy office where he closed the door, assured me that it was 'absolument pas nécessaire' to identify himself and chugged his third bottle of my precious beer as he lingered over my letter of authority from the National Parks. Oozing sweat and menace, he was the sinister embodiment of a Western intelligence report which wrote 'SNIP exercises almost unchecked powers of arrest, imprisonment, and interrogation. It has used these powers to intimidate individuals or groups posing a real or imagined challenge to the regime's authority.' He certainly intimidated me for most of that afternoon, although there was a reassuring sentence in that report to which I knew we would eventually get round: 'Much of their resources being the product of extortions and theft'. As a foreigner with an aircraft, a motorbike, binoculars and a big camera – from his perspective unconvincingly posing as a simple birdwatcher – he told me that I was a clear and present danger to the regime, and a threat to the region's stability. He could, and really should, lock me up. However, a generous contribution to SNIP's resources would nullify all that, and if I could provide another round of beer my papers would be stamped and I would be free to watch as many of Zaire's '*belles oiseaux*' as I wished.

I spent several hours, and a lot of dollars, with the man from SNIP, the frightening fact being that there was no alternative – out here he was the law. Once we were done I walked back through the darkened village to Père Léon's house, from which floated the sounds of young laughter and scratchy Beethoven coming from an ancient wind-up gramophone. The main room was full of dancing children, their bodies gleaming and eyes shining in the fitful light of a couple of smoky oil lamps. There I found Léon and Giles sitting across a grubby five-litre plastic jerrican half-full of palm wine. 'Welcome to Léon's Disco!' called Giles, pouring me a glass of the opaque liquid and pushing over a low chair made of three sticks and some antelope skin. The routine in this house was that every morning a single shotgun shell was given to a lad of about twelve, the eldest of the six orphans who lived with Léon, with which he was expected to provide dinner – usually a monkey, occasionally a duiker antelope or a forest guinea fowl or, once or twice a week, nothing. That night we all sat at a long table and while Frankie Lane belted out 'Ghost Riders in the Sky' we tucked into a fine colobus monkey stew with manioc greens and fried bananas, washed down with plenty of fresh palm wine.

Later, when the children were sent off to bed in a dormitory at the end of the building, I asked Père Léon, over a cup of the bitter local coffee, about the many pictures of men and women on his walls. As the Beatles sang 'Let it Be' he took me on a heartbreaking tour of those portraits, one or two smiling, but most of them stiff and earnest in a sort of Congo Gothic. All missionaries or nuns, and dear friends of Léon, they had been tortured, raped and killed in Stanleyville, a hundred miles to the west on the Congo river, in the bloodshed that followed Zaire's independence in 1964.

'I should have been with them,' he said quietly, almost regretfully. 'At the time there were two of us here at Opiengé. Armed rebels came and took us to Stanleyville where we were locked up with some other white people. We had only just arrived when the Belgian paras landed in the city on the left bank and we were soon freed.' He paused, the thirty-year-old memories obviously still vivid and painful to this gentle soul. 'But they did not get across the river in time, and every person whose picture you see here was horribly murdered. It was the 24th of November . . . I should have been with them.'

As I put my arm round his shaking shoulders I wondered at the strength of his commitment to a people and country that had brought him such pain. He had been lucky. The news in the sixties had been full of the Congo's horrors and I well remembered reports of the paratroops landing at dawn and racing through the city. Many of the hostages had been taken out and made to sit in rows in the square, and as the paras moved in, the rebels had opened fire on the innocents. When the troops arrived just minutes later they had found over eighty men, women and children dead, dying or wounded. That massacre had involved the slaughter of whites, so had been one of the most publicised atrocities of that awful time, but many other massacres – infinitely more barbaric and involving thousands of Congolese – took place across the vastness of that benighted country. Their cries lost among the great trees, their blood washed away by the incessant rains, out of sight, undocumented, forgotten.

The next day I invited Père Léon to come with us when we took off in the Cessna for an aerial recce of the area we were about to tackle on the ground. To make room for the baggage, I had taken out all the seats except mine, and Giles and Léon had to kneel on the floor in order to see out of the windows. Léon had not flown for decades, and never so low over the forest and serpentine rivers. Looking over my shoulder, I called out that I hoped it wasn't too uncomfortable. 'I've spent much of my life on my knees,' Léon shouted happily, 'but never with such a wonderful view.'

We still had a long way to go to where I wanted to concentrate my search, so two days later we said au revoir to Léon and steered our heavily laden bikes down the footpath towards the Loya river and the Maiko National Park, some four thousand square miles in the centre of the most remote part of the eastern Congo basin. The only traces left of a once major road were a couple of iron bridges from which the wooden floorboards had rotted or been stripped. Crossing the bridge involved walking the tightrope along a narrow metal beam while reaching across to guide the bike along a parallel beam some three feet away. It was exhausting work, but far less frightening than one river crossing we came to which was just a single slender tree trunk. Giles had been biking for years, and with experience and the wildness of youth he took it at speed with a banshee yell. I rode very slowly across, keeping the bike upright with my feet, trying not to look down at the rocks and foaming water ten feet below.

At Balobé, the last tiny village on the track before the Loya river, we recruited some porters and guides, left our bikes and started the long hike to the Jangua hills, deep inside the park. As we walked the porters sang their traditional antiphonal carrying songs, and I knew that Stanley and Chapin had heard these same rhythms as they marched into the unknown. At the river we hired a dugout canoe and soon we had all made it across and into the dense forest. For the next day and a half we marched through high forest, my calves burning and shoulders aching under the weight of my cameras. Under a threatening sky we came to the Jangua river and made camp close to a foaming waterfall under which we all showered. Giles and I quickly put up our tiny one-man tents and helped the crew to collect firewood and build a shelter in which they would eat and sleep, a simple framework thatched with large *mangongo* leaves, an arrowroot that the Mbuti use for their houses.

The next day we fanned out into the forest looking for signs of the peacock. It had rained in the night; monkeys were hunched in the dripping trees and a watery sun lit wisps of mist that drifted like ghosts between the trunks. Game birds are fond of bathing in dust or damp soil and their shuffling creates distinctive shallow pits that often contain a feather or two. We soon found several such bathing sites from which we collected two plumed guinea fowl feathers with their distinctive spots. Examining the hollows, I felt that a couple of them were larger than one would expect for guinea fowl, so I set up a hide some distance away from which I could film if the visits continued. Below the giant Mbeli trees we found distinctive chicken-like droppings and scratchings where whatever these birds were had searched the leaf litter for the dark purple olive-like drupes. The best – and often the only – way to film a shy bird that lives in dense vegetation is to find a nest and this was our top priority. Many forest trees have buttress roots, blade-like projections that arch down to the ground all round the trunk, so the tree resembles a multi-finned rocket. We searched the deep, dark indentations between the buttresses, hidden places where over the years I had found the eggs of rails and forest francolins. Every day our hearts would pound with excitement at the characteristic explosive take-off of a large game bird, but it was always just a plumed guinea fowl.

The forest was rich with larger creatures, and within a week we had seen the spoor of leopard, elephant and buffalo, and the smaller tracks of red river hogs and various antelope. High in the trees we saw the nests of chimpanzees, roughly woven with living branches, and twice we encountered the large dung-filled ground nests of gorillas. We also found two small chestnut feathers with iridescent green tips that could only have come from a peahen – we were on the right track. Every evening, after a refreshing shower in the waterfall, we climbed one of the nearby hills and sat watching the sunset in the hope of hearing the peacocks which, we'd been told, call loudly on going to bed. There was one spot from which we could see out over range after range of softly rounded hills, the trees on the ridge tops in dark silhouette and the valleys filling with luminous mist as the air cooled. As we sat there, the evening calls of colobus monkeys rumbled from one valley to be answered by those in the next and I could imagine that wave of sound travelling right across Africa as the terminator, the grey line that moves across the globe bringing nightfall, made its way west. In the distance a family of chimpanzees were preparing noisily for bed, groups of great blue touracos burst into their manic kurrraaaa, kurrraaas, crowned eagles whistled high overhead and afep pigeons cooed incessantly. And although there were no unidentified cries in that soundtrack to bring us quickly to our feet, every evening was a joy.

Then, one memorable morning I was resting on a tree stump in the forest watching and listening. After a few quiet minutes I heard movement in the leaf litter. I had been watching a squirrel that had moved out of sight, and I assumed it had jumped to the ground and was searching among the leaves. I swung my binoculars in the direction of the sound, coming from a fallen trunk about twenty yards away. Suddenly the electric-blue head, red neck patch and stiff black-and-white crest of a peacock appeared over the top of the log. I froze, my heartbeat registering in the jogging of my binoculars. Then the bird lowered its head from sight and walked out from cover. It was the size of a small turkey hen, plump and shimmering. For maybe ten seconds the dappled sunshine lit the metallic lustre of its blue breast and wings, and its emerald-green back, before it walked slowly into a dense patch of ferns and was gone. I stayed a while to see if it would reappear, but couldn't wait to share my excitement and raced down the hill to find Giles.

I was still glowing two days later, but had to face the reality that this was my last day there. Giles and the crew would stay behind to continue the search, but I had to leave as I had been asked to judge the finalists at Wildscreen, an important British wildlife film festival, and would have to start walking and biking out early the next morning. It was bad timing, but I told Giles that if he had no luck finding a peacock's nest or a regular feeding tree where I might be able to film the birds he should try to capture one. The next morning, with a porter carrying my camera boxes, I set off on the twelve-hour walk back to Balobé, where I'd left my bike. I was heading towards the other half of my life. And that's when I got trapped under my bike wondering what the hell I was doing there . . .

Days later I was sitting at my table at the Wildscreen awards ceremony dinner, my hair cut, my nails cleaned, fancy waistcoat, the works. Jenny had come to the festival with me for a couple of days, but had been too tired to join any of the screenings or discussions that punctuated the events, and had left for Manila, where she was visiting more healers, looking frail and frightened. Before she departed, one of my fellow film-makers – a tough Aussie with a soft centre – had passed around a notebook in which everyone had written heart-warming messages of support. I was among a very special and loving community of like-minded friends who were a great help getting me through those days before I could join her.

On the night of the awards I had done my dancing bear routine for the evening, having presented one of the awards, and felt it was now safe to relax and have a stiff drink. My old friend David Attenborough came and sat with me for a while before going up to the stage to do his own party piece, presenting the Lifetime Achievement Award. I had been lobbying for my old mentor, Des Bartlett, and hoped it might be him, but well into my second Rusty Nail, I was only half-listening to David's familiar voice when I heard him say: 'If you are setting up a shot in Africa and you are about to say, "and here, on the plains of . . ." and a plane suddenly swoops overhead, pruning the thorn tree you are standing under and ruining your shot, then you know that is the recipient of this award.'

Oh hell, he's talking about me! I had been on the verge of tears for two days and now I would have to make a coherent acceptance speech. As David rabbited on about what a fine, adrenalin-addicted fellow I was, I felt like nothing of the sort and frantically scribbled some names and milestones on the back of the menu and emptied the water jug.

'And Alan, almost single-handedly in my opinion, made wildlife films grow up, to become the well-crafted and professional productions that we know today.'

Well, thanks, Dave, I thought, but keep going. I need more time, more water. As he ended, I sleepwalked up to the stage, leaving my notes on the table. In the event the Rusty Nails kicked in and I gave probably the best and most sincere speech of my life.

'I have been lucky to have had the love of two extraordinary women in my life. My Joan, who helped so much to build my career, and my Jenny, who has helped me to see that a career is one of the less important things in life. So thank you, Joan, thank you, Jenny, and thank you all those friends who have walked along with me on the big safari.' To fight back the tears I switched to humour and got through to the end. 'I'm not retiring. I'm not being put out to grass. I'm certainly not being put out to stud – dammit! So, in the eternal words of television: right after this break, I'll be back.'

I staggered to my table, ears ringing and eyes burning. I wished there had been some way that Joan could have shared the honours – she deserved them – and yes, I would be back, but I wondered if Jenny would be with me.

The next night, in one of those bewildering dislocations of international travel, I was in the Philippines. Normally, this would have meant I was on my way to film the tarsier, the tiny, bug-eyed primate that still hung on in small pockets of forest, or the huge monkey-eating eagle. It was unsettling to arrive in a tropical country without wildlife on the agenda.

Jenny had been joined here by her best friend Carol Byrne – a serene Kenyan lady who had been a pillar of strength for me in those painful years. Jenny had arranged an appointment in Manila with Alex Orbito, a healer of world renown who had treated kings and beggars alike in many countries, with a large percentage of his patients claiming his cures worked. He modestly said he was merely an instrument of the Divine and apparently refused to acknowledge any personal praise. He did, however, acknowledge payment. My confidence in him was severely undermined when I discovered that one of the things that had brought him to world attention was a book by the film star Shirley MacLaine, in which she described her experiences with him. Shirley was, after all, a woman well known for the absolute certitude with which she described the various people she had been in previous incarnations. I couldn't put out of my mind a cartoon I had once seen of two iguanas in the tree-tops with one saying, 'It's funny, I keep thinking I used to be Shirley MacLaine.' She had also written about a visit she had made to a Maasai village, filled with torn and bloodied warriors fresh from spearing a lion. Anyone who spent any time in Kenya would have recognised this village as a well-worn tourist spot and her story as a load of absolute cobblers. So for me, Alex didn't come with the best of references.

On arriving in Manila I discovered that Jenny had flown to an inland town called Bagio to see a man named Placido, one of Orbito's disciples. By the time I got there Jen was utterly exhausted, and Carol and I feared for her life. Placido's consulting room was a short taxi ride out of town: a single small room attached to a garage that, judging by the noise, specialised in panel beating. Placido was short, slim and thirtyish, with darting eyes and a pleasant smile. When Jenny climbed on to the high, narrow bed in his room, Placido pulled up her shirt, revealing her taut bulging tummy, the result of her liver and spleen being hugely engorged with abnormal white blood cells. After resting his hands on her for some minutes, he slowly pushed the fingers of both hands deep into the flesh by her navel, keeping them there while he whispered some incantation. Slowly, blood began to fill the depression caused by his fingers, spilling out and running down her side on to the sheet – not a lot, maybe an egg-cup full. After some time he removed his fingers and from between them took some pieces of what looked like liver. This, he declared, was 'what is making her sick' and he dropped the pieces into a wastebasket before washing his hands.

Jenny was totally drained by the experience and it was some time before she felt well enough to head back to the hotel. I didn't know what to make of it. Placido – I *must* stop calling him Placebo! – had welcomed me to stand as close as I wished to watch him. On one level I had been moved and entranced, but I also knew that what I had been watching could have been some rather elementary sleight of hand. Instead of scraps of liver, an expert could equally convincingly have produced the coloured scarves, doves or rabbits that are a magician's stock-in-trade. But by this stage Jenny and I had visited so many healers and gurus that I no longer knew what to believe. And then again it didn't matter. Jenny believed and one thing I *did* believe was that a strong enough belief in the efficacy of something can make it work for the believer. Perhaps Placebo wasn't such a bad name for him after all.

We flew back to England, where Jenny slowly regained her strength in the house in the Cotswolds that was my base camp when editing my films. Looking out over the soft hills and tiny streams, I ached to get back to the deep forest and great rivers of the Congo. Jenny had recently developed a great urge to run her fingers through rich soil. It calmed her, and wherever we were I would fill a small bowl and she would spend hours kneading it therapeutically. One evening, after a heavy storm, I went outside to fill the bowl from under the eaves where the soil would not be too wet. Coming back in the dark I slipped and fell heavily. Face down in the mud I was suddenly back on that trail in the Congo, my mind full of questions. Digging my fingers deep into the mire, I sobbed in despair. What the hell am I doing here?

Before leaving Giles in the Maiko forest looking for the peacock, I had suggested that the best way for him to capture one would be to use a snare and I had designed one that would not harm the bird. Snares are usually attached to a sapling, which springs upwards, tightening the noose and suspending the bird, often dislocating a leg. I had designed a kinder version with an elastic connection between snare and sapling, and a check cord that stopped the sapling springing too high. The snare would hold firmly, but with some give in it, and would not lift the bird off the ground. Giles and his team had persevered with these and just before I got back to Epulu from England they had caught a male peacock – a lovely specimen he named Napoleon.

I flew straight down to Opiengé, taking the wine, chocolate, cheese and salami I had brought in duty-free for Père Léon, and some cold beers to toast Giles's success. Napoleon settled down well in a large enclosure at Epulu, while Giles and his team geared up to find him a mate. Almost a month later they caught Josephine. She was arguably even more beautiful than the cock, her head, neck and underbelly a rich tawny orange, with an emerald-green back and chestnut wings barred and spotted with bronze and violet. They made a stunning pair, but sadly, being rare and beautiful did not mean they were interesting. (If only some of our 'celebrities' would learn that.) All they ever did was strut around, eat tiny amounts and endlessly preen – recognise anybody? I guess I shouldn't have been disappointed. Despite their legendary status they were, after all, just glorified chickens. From the filming point of view it was small reward for such a demanding project, but the real prize had been twofold. In finding and seeing this elusive bird in the wild we had opened a curtain that had been closed for decades, and for me personally there was the satisfaction of adding the final item to that sailing-ship cargo of boyhood memory. Ivory and apes and, now, finally, peacocks.

Back at my Epulu base I took stock. I had now filmed many fascinating creatures including some of the rarities I had set my sights on – the okapi, the giant otter shrew, the chevrotain and the peacock – all of which had been at least seen by naturalists at some time in the past. But the last animal on my list was a total mystery and almost no more was known about it than when Chapin and Lang had collected a couple of skins back in 1910. This deeply secretive creature was the aquatic genet, *Osbornictis piscivora*, known to the Mbuti, by name only, as *apakékéké*. For years I had been showing Mbuti friends pictures of the creature, but only one old man thought he had seen something like it, drowned in his fish trap, but he wasn't sure. Early in the project I had made a trip to the museum at Tervuren, in Belgium, to see the few rather tatty *Osbornictis* specimens they possessed, the ragged skins of animals caught in gin traps, or killed by the hunters' dogs. I realised as soon as I saw them that, despite its Latin name of fish eater, this animal had not evolved to pursue fish in the water like an otter. It was built more like a marsh mongoose and I felt I would find it hunting along the edges of marsh or shallow streams.

Over the years I had often visited a beautiful area of Mbau forest some ten miles east of Epulu. Winding through this ancient forest were many such streams, with golden sandbanks that every morning spelled out the night's activities. Along with the slots of antelope were shards of mussel shell and long splayed fingermarks that told of a marsh mongoose, the short fat, clawless digits of the Congo swamp otter – and occasionally a trail that raised my pulse. Some small predator with tracks like a common genet had spent time walking along the stream edge, with frequent stops and forays into the water. I had tucked these memories away and now, together with Giles – who was staying on to help with this final challenge – made a recce of the area and worked out where to set the live-catching traps and how we would operate. He would bring fresh fish for bait every evening on his motorbike and an Mbuti would race to bait the traps, walking in the stream so as to minimise the human scent. At dawn the next morning he would revisit the traps, and if anything was caught would cover it with a blanket and bring it to the main track, to be collected by Giles. I rode out every day with Giles for the first week, but it was obviously going to take time, so I decided to get on with other filming while he made these daily trips.

One day Giles came roaring through the village, standing up on the footrests and yelling, '*Apakékéké iko!*' ('The genet is here!') He raced into the house with a huge grin and I opened the blanket with trembling hands to find a dark-brown marsh mongoose. Another time I came in from filming to be handed a celebratory mug of wine and be told, in the Maurice Chevalier French we had adopted when matters demanded a more serious language than English, '*Mon général, votre animal est sous le lit!*' I dropped to the floor, took an excited look and replied, '*Merci, mon brave, pour un autre mangouste de la swamp.*' Even though he had been showing a colour picture of the genet to all and sundry for weeks, for some reason he seemed unable to see, in the flesh, what was clearly *not* the animal in the picture. But Giles was loved and admired by all the village. He would do wheelies and jump the ditches around their huts, long hair and fringed buckskin jacket flying. He was known as '*le style*' or Tonton, not for the cartoon character but for the Africans' attempt to say Thornton, and his mannerisms were much copied by the young village bucks.

Over the next couple of weeks we moved the traps many times, then one day I again heard the cry '*Apakékéké iko!*' and saw Giles punching the air as he did a second circuit of the village to a noisy welcome. I felt sure that this time he must have it right and sure enough, removing the blanket, he unveiled a beautiful adult female *Osbornictis*, unharmed and unruffled by all the noise. We sat for a long time just feeling good and drinking in what no naturalist's eyes had ever seen before. I was reminded of Alfred Russell Wallace who, a hundred and fifty years ago, had come up with a theory of evolution by natural selection similar to Darwin's. When he collected the first specimen of the exquisite king bird of paradise he had written, 'The emotions excited in the mind of a naturalist, who has long desired to see the actual thing which he has hitherto known only by description, drawing, or badly preserved external covering – especially when the thing is of surpassing beauty and rarity – require the poetic faculty fully to express them.' I lacked his poetic faculty, but my mind was certainly full of excited emotions as I gazed at this thing of beauty and rarity. We called her Dorothy, in celebration of a lady in Giles's interesting past.

Dot's dense fur was a bright rufous over most of her body, with a jet-black bushy tail, legs and muzzle, and chalk-white throat and cheeks that sprouted long, thick, downward-sloping whiskers. She was so calm that I feared she was in shock, but when we caught a male two weeks later, he too was totally unfazed, and we soon saw that slow and calm is the way that aquatic genets go about life. I soon moved them into a large enclosure with a small stream and a string of pools running through it, and we settled down to watch them closely.

The chevrotain had revealed unimagined behaviour with its underwater walking, and soon the genets gave me a similarly unexpected surprise. The male was the bolder of the two and more richly coloured, and he became the star performer. Creeping very slowly along the stream edge, he stopped and reached a front foot out over the water, held it there with the wrist bent sharply upwards, then quickly gave the surface two or three gentle pats. Fly fisherman have a similar technique of bouncing the fly on the surface, called dapping. Putting his head forward, he then lowered his whiskers on to the surface, where I felt sure he was reading vibrations from fish either disturbed or attracted by the patting – perhaps thinking it was an insect hitting the water. He repeated this exercise all along the bank – which explained the sandbar tracks I had seen long ago – then stopped. Raising his head, he tensed his body, then struck at high speed with open mouth, his head completely submerged. Emerging with a fish caught firmly between his teeth, he raced off to cover. This was the only time we saw them move really fast and was behaviour probably designed to avoid the attentions of marsh mongooses, which might well rob them. This behaviour was such a gift, for not only were the *apakékéké* matchlessly rare, they were extremely handsome, and in their calm fashion they showed us photogenic behaviour that film-makers can only dream about.

There was another great bonus too. For years the Mbuti had been showing me things I would never have seen myself, sharing their knowledge of the forest plants and animals. Now I was able to show them something from their own forest that they had never seen, and enjoy their entranced expressions and shaking heads as they watched the genets fishing. Much as when we had shown the Maasai that flamingos hatch from eggs, this moment with the Mbuti was incredibly precious, a unique experience that I treasure as much as seeing the animal for myself. The genets got on well together and for a while I considered waiting to see if they would breed, but in the end decided that they should be allowed their freedom. We released them back on to the same stretch of river, and their tracks across the sandbank and into the forest triggered memories and a happy smile.

All the weird and wonderful animals that I had managed to film over the years in Zaire featured in the last of my films – *A Space in the Heart of Africa*. Among other things, this film showed how fundamental elephants are to maintaining diversity – and how, by their feeding habits, they slow down regeneration in tree-fall gaps and so provide reachable fodder for many smaller species. My other films had been about the Hart family and their work with the Mbuti and okapi, the Virunga National Park, from the glaciers of the Mountains of the Moon to the erupting volcanoes and gorillas at the southern end, and the story of Garamba and its elephants. Mission impossible had been accomplished and, thanks to vet Karin's devoted care, all our captive animals had been successfully returned to the forest. Thanks, too, to my invaluable ragtag band of helpers who, as we feasted together one last time on roasted goat and lots of beer, divvied up the furniture, bedding, petrol drums and many other items I was leaving. Rogé was still with me, and he received my solar panels and lights while my faithful cook Vomit, her petticoat finally worn out and looking regal in the new dress I had brought her, claimed her stove and pots and pans. I donated my house and Land Rover to the National Parks, and as I flew out I knew my airstrip would revert to cultivation or forest.

With a lump in my throat I lifted off out of Epulu for the last time, with only some aerial photography to do before my operation was over. I flew over the wide brown rivers that snaked and looped across that ocean of green, the big broccoli, thankful that I had survived its storms and featureless expanse. I had my usual fearful thoughts of carbon dioxide as I filmed the smoking crater of Nyamulagira, and froze as I flew along the great snow-covered spine of the Mountains of the Moon, its glaciers dripping water destined for the Nile. I landed at Goma to say goodbye to the National Parks people there and the next morning I was up at dawn to fly to Nairobi with just some filming to complete over the highland forests of the Virunga volcanoes. It was a sublime day, with the crater lakes glinting in the sunshine, the jagged fang of Mikeno wreathed in mist and a sprinkling of snow on Karisimbi. In the saddle between those great peaks nestled the clearing and remains of the hut where Joan and I had first filmed gorillas, and introduced them to Dian Fossey, and to the east was the small research station that Dian had built at Karisoke. I'd got my aerial shots, and was daydreaming, flying just for the joy of being high and free over those beautiful, memory-filled mountains after years in the forest, where a long view meant forty yards and you never saw a horizon.

I was well into Rwandan territory, a country with a good air force, but also a peaceful nation in which I knew no one would bother much about watching the skies. I turned on the radio, hoping for some music to my match my mood, but instead a BBC news flash came on and in seconds the world had changed. A plane bringing the presidents of both Rwanda and Burundi into Kigale had been shot down during the night, with no survivors – just a few hours ago. All Rwandan borders had been closed, and there were reports of explosions and heavy fighting around the presidential palace in the capital – no more than fifty miles away, merely minutes for the jets that must have been scrambled and be nervously patrolling somewhere out there. I put the nose down and pushed the throttle to the wall.

Rwanda and Zaire were about to enter yet another tragic cycle. Within hours the tribal killings had started and in the next three months eight hundred thousand people were slaughtered, mostly with machetes or hoes, and two million fled across the border to Zaire, where their vast refugee camps destroyed great swathes of the Virunga National Park. Soon Dian's research station would be burned down and those serene forests would contain many more guerrillas than gorillas, whose presence in Zaire led to years of conflict, while the world's largest peacekeeping force looked on impotently at the rape, death and displacement of up to five million innocents. I watched my GPS as the little aircraft icon crept across the screen, and finally over the border and out of Rwanda. I circled for a last look at Zaire, the beautiful, blood-soaked country that was burned into my spirit. The stately row of volcanoes, the sun flaring off the lakes, the long, distant trace of white in the sky that was the snowcaps of the Mountains of the Moon and beyond that the shadowed edge of the forest that led to Conrad's heart of darkness.

Then I booted the right rudder and swung to the east. I was going home.

11

Requiem

Jenny was losing weight and energy, but not spirit. Back from the forests, I was glad, now, to be able to give her more time. She came to England while I was editing the Zaire films and when I delivered the last of them I announced I was folding my tent and tripod, and giving up film-making. The films I made meant spending long periods of time away from home and I didn't want to do that any more. The films had done well, but television was changing. David Attenborough was still doing his brilliant series – thank you, Lord, for that wonderful man – but the trend was now towards quick and cheap productions featuring charismatic presenters. Many of these films showed that animals – especially snakes, crocodiles and anything poisonous – were something you picked up, basically molested and used to augment your ego. This genre taught viewers almost nothing about the animal and its unique way of life, and I believe did great damage to a generation of children who grew up learning these attitudes towards wildlife. All they understood was that their brave presenter was a hero – who warned them not to do this at home – and, of course, in a world where millions of people live risk-averse lives, these films were very popular. I was glad to be out of it and though I missed my work, in caring for Jenny I had a purpose. When all my cameras were stolen one day I took it almost as a blessing, at least now when I opened their cupboard all those lenses no longer gazed at me pleadingly, like eager dogs hoping for a walk: it made retiring easier.

Twelve years into her illness, Jenny wanted to visit some of the wild places she loved and see them for what she knew would be the last time. Her body could not take the pounding on Kenya's notoriously awful roads, so I did something I had long fantasised about and bought a small helicopter. It would make the journeys possible, was a fresh challenge and would provide the adrenalin rush that my new life lacked. And this certainly turned out to be the case. They say you can't teach an old dog new tricks. I was quite an aged hound and a helicopter is a very tricky new trick. Except that I was comfortable defying gravity, my four thousand hours flying fix-winged aircraft and hundreds of hours in balloons were no help; flying a chopper was a whole new art. I quickly crashed and wrote off two, bought a third and finally got the hang of it.

I could now take Jenny in comfort on farewell trips to the Serengeti, the bush country of Tsavo, Lake Baringo and many other places that she loved. She was completely dismissive of her obvious decline, always laughing despite her pain, and I realised how much I had come to love this courageous and sunny lady. But she was growing gaunt, and when she became too weak to handle the heavy clay and shape her pots, I felt for the first time that she was giving up.

She wanted to have Christmas in England, then go for some treatment at the Park Attwood Clinic there. Soon after our arrival I realised this was actually a hospice that cared for the terminally ill and felt sure that Jenny was doing this to relieve me of the pain of being her main carer. It was a wonderfully serene place where I was looked after as well as she was. I could have counselling to help me deal with the coming loss, a massage to take the tension out of my shoulders, and the use of a nearby gym and pool. Our lovely friend Carol Byrne joined us there, bringing as always her aura of deep spiritual peace, and I spent the next couple of weeks coming to terms with what was looming. I left Jenny for just one night to go up to London for the Millennium New Year's Eve celebrations. I felt strangely disconnected standing looking up at Big Ben, waiting for midnight, jostled by millions of people celebrating and welcoming the future, knowing that Jenny's was measured in days and mine so uncertain. Ten days later she died in my arms and for the first time in years I could hug that pain-racked body without fear of hurting her. Back in Kenya, I placed her ashes in one of her elegant pots and buried it under her favourite tree: the end, for both of us, of a long, harrowing, loving journey.

Over the years Joan and I had remained good friends, but time and circumstances had moved us so far apart that we now found conversation difficult. We were both very different people from the couple who, twelve years earlier, had thought their long safari together was only on hold for a while. I knew that Joan had kept a candle burning for me for several years and it had saddened me that for someone so talented she had for so long seemed to live in my shadow, but that had all changed. While I had been buried away deep in Zaire, Joan had been flying. She had played an important role as spotter and photographer in a major aerial count of hippos in Zaire and elephants in the Central African Republic. She took trips to Antarctica and the Kalahari, and joined expeditions in search of rock art in the wildest parts of North Africa. Joan was very much her own woman now, joining clubs and giving dinner parties – and she even took a lover for a while, which I hoped would bring her happiness and perhaps a solution. But in the end Joan's solution came not from her love of someone new, but her bone-deep love of wildlife and her land – Africa.

When we bought our eighty acres on Lake Naivasha's south shore there had been only a few isolated houses sitting on large acreages, surrounded by wildlife. Herds of zebra and impala came down to the lake to drink, giraffe browsed the forests of acacias along the shore, and buffalo and hippo came out of the reed beds at night. Now, Joan's patch was one of the last few refuges for wildlife along that shore, which huge flower farms were covering with hundreds of acres of plastic greenhouses. Naivasha had rich volcanic soil, plenty of sunshine, a lake full of fresh water and even geothermal energy on tap. When I had lived there in the seventies a couple of local farmers had started growing peppers and courgettes for export, but flowers had raised the bar to an astonishing level. Commercially, the flower business was a great success story. Started by a handful of visionaries, it had grown to a two-hundred-and-fifty-million-dollar industry, exporting half a billion blossoms a year, most of them from Naivasha.

But the commercial success came at a great social and environmental cost, to the extent that a renowned freshwater biologist wrote, 'Without any kind of active ecosystem management, supported by all the diverse stakeholder interests, Lake Naivasha will soon change beyond all recognition, becoming Africa's own Aral Sea.' One reason for the flower farms' success was cheap labour; the industry needed vast numbers of workers and from every corner of Kenya they came in their tens of thousands. Naivasha town, once a sleepy little place of dust, dogs and donkeys, had exploded into a festering slum, housing – if that is the word – a quarter of a million indigents, most of them unemployed. Satellite slums spread round the lake, scruffy counterpoints to the acres of gleaming hot houses that now lined the shores. The run-off of effluent from the slums and chemicals from the farms was turning the lake into a giant sewage pond, riddled with toxins, and the thousands of frustrated work seekers were turning to poaching fish and wildlife, and to crime. As so much of the lake shore was now fenced off, Joan's open land became the easy way for fish poachers to access the lake. Although she did not want to block the passage of wildlife down to the water, the trespassing turned so serious that she finally had to erect a fence; but, as they were doing to so many other fences, the poachers simply cut the wire and used it as snares to strangle zebra and impala.

For a long time Joan tried to chase off the men who brazenly set their undersized nets right in front of her house. Here in the shallows they caught all the hatchlings that should have been the next generation, and soon Joan was leading a one-woman campaign against everything and everyone that was destroying the lake. She employed men to chase the poachers, financing motorboats, radios, mobile phones and vehicles, and personally delivering captured poachers to the police. She then tried a new tack, taking some fish poachers and turning them into legal fishermen by buying them nets, fishing licences and even a boat. They were now legally self-employed, with powerful personal reasons to apprehend the illegal fishermen who were ruining their new way of life. These men morphed into a task force that was soon cutting a swathe through the poachers. Joan was on call, day and night, to come and collect captured poachers and haul them off to the police. Some of these men were badly beaten by her team of vigilantes – and she was obviously unable to restrain them. For a woman living on her own she was being far too hands-on and visible.

Almost any illegal activity in Kenya ends up being tithed or, if it is highly profitable, taken over by the very people whose job it is to curb that activity. Thus Joan made enemies not only of the poachers who were beaten or imprisoned by the authorities, but finally also of the police and fisheries officers who were losing out on bribes or protection money thanks to her zeal. Almost every member of the task force and Joan's staff had talked her into advancing enormous loans to them that they could never have repaid. Everyone was taking advantage of her and the whole operation was collapsing. Many of the people close to her, myself included, encouraged Joan to persuade the Riparian Association – which represented the owners of land that bordered the lake – to take over the running of the task force. But Joan continued to finance it, and to buy plots of land, motorbikes and mobile phones for a whole range of scroungers who were meant to be helping her. When the task force was finally closed down, and she had to lay off the team, they vilified her for taking away their livelihood. The whole exercise had been thankless and exhausting, and soon the illegal fishermen returned to operate in front of her, and every other, lake-shore house, her animals were snared and the hangers-on became more demanding and intimidating. Joan had been transformed into a prey animal.

Joan's friends had grown increasingly worried about her security. Two white farmers and a tour operator had been murdered within a few miles of her home, and she had been hijacked at her gate and was lucky not to have been hurt. Fear was closing in on Naivasha, but when I suggested calling on her at home her reply was fierce:

'You would hate it now! It's all bloody plastic. Polythene hothouses and polythene bags. That lovely stretch from the main road to the house is all speed bumps to protect the squalor that presses against the road. Just poverty and thuggery where you remember only wildness and beauty. You would hate it! Hate it!' Then she listed everything else I would hate: the latest robberies, attacks on farmers and poaching that were now Naivasha's quotidian. I asked her, as always, whether she was safe. 'All I do now is finance the operations and keep a low profile, but everyone knows – especially the baddies – that I am still the point man. I know I am marked, but I've reinforced my house, I have security backup, and I will carry on' – which is about as far as I or any of her friends ever got with that subject.

Less than a month later, on 13 January 2006, the phone rang

at two in the morning, and my friend Adrian Luckhurst told me that Joan had been shot and killed in her bedroom. As always, the line was bad and he didn't have much more information, so I simply said I would pick him up with the helicopter on his lawn at first light. I made some tea and sat trembling, trying to stop my heart from bursting with anger and hatred. I raged and cursed Africa and Africans, this wild and brutal continent that I loved so much, but which exacted such a premium in exchange for its joys. I wanted only to find whoever it was and shoot them – hunt them down and kill them as mercilessly and as savagely as they had killed her. Yet deep down I knew that for all my thoughts of revenge and justice, I would get neither in this corrupt and lawless society. I would not poison my life with an endless quest for justice that I would never find. I would *not* be like poor Julie Ward's father, who had spent more than a decade and most of his fortune trying to find the killers of his daughter in the Mara Game Reserve. He had been encouraged, then marginalised, given hope time and again, then crushed under the uncaring weight of ineptitude and corruption. I was not going to launch a crusade inevitably doomed to heartache and failure. That decision calmed me somewhat and I went out into the clear, cold night. 'Don't take your anger into the air' is a profound piece of pilot wisdom, and I sat in the chopper shivering and sorting out my thoughts. As soon as I could see the horizon I took a deep breath, lifted off and headed for Naivasha.

Flying is a form of therapy for me at all times and it was especially so on that dreadful morning. Back in the sixties and seventies I had flown this route hundreds of times with Joan in our little two-seater, the slow climb to the north-west over the Kikuyu country, with its small, well-kept farms wreathed in the smoke of morning fires; then the bit we had always enjoyed most: flying low and fast, and suddenly having the world drop two thousand feet from under us as we crossed the steep edge of the Rift Valley escarpment; then down, down over the plains with herds of zebra and gazelle, and round the deeply fissured flank of the extinct volcano Longonot. On those flights the shoreline that then came into view had been almost all woodland, stands of fever trees with their lemon-yellow bark and giant euphorbias looking like the candelabras of their Latin name. Now the shore was just as Joan had said – wrapped in polythene – and she was right, I hated it. For mile after mile the shoreline was lined with plastic greenhouses that from the air looked like gigantic plasters on the gashes of a wounded land. Parallel to them were the strung-out slums where tens of thousands eked out a desperate and resentful living sustained by futile hope and the offerings of the poachers. But there were still a couple of greenish gaps, and in one of them stood the tallest trees on the lake, a group of ancient, giant eucalyptus that grew right behind our house and had always been our homing beacon.

I had not been into this house for twenty years and I choked on the memories as I was suddenly surrounded by reminders of the life Joan and I had shared: the stone axes from New Guinea framing the fireplace; the mounted flamingo chick shackled with its soda anklets; the print of the balloon sailing over Kilimanjaro. The memories all came cascading back, the safaris and camps, the many creatures and wonderful people of all races who had been part of our journey, the triumphs and failures and the real love that had somehow gone wrong. 'Oh Joan,' I sighed, steeling myself to go into the bedroom.

The bleakness of that room was a terrible shock. I'd had no idea of the extent to which Joan's life had been narrowed down. For all her adventurous trips to the desert and Antarctica, she came home to a desperate and fearful place. The heavy steel sliding door to her bedroom had clanged shut every evening, leaving her with a bed, a reclining chair, a small TV and two large boxes of videos, almost all wildlife shows, many accompanied by copious notes on their content. I could not imagine Joan in this grim prison – Joan, who had sat by thousands of campfires, slept under the stars and tried to watch every full moon rise, locked into a giant safe watching old films. I was immeasurably saddened and wondered how many people – if any – realised how her battles with lawlessness had rewarded her. The forlorn features of that room made their deep impression before the rest swam into horrific focus.

There was a window on either side of the bed and both of them were riddled with bullet holes that had been fired from so close to the glass that the muzzle blast had set fire to the curtains. The bullets had sprayed the room, making ugly holes in the walls and splintering the furniture. The blood trail clearly told the story and was the easiest – and hardest – trail I ever followed. Hit by the first burst and bleeding, Joan had headed for the bathroom which, with a heavy steel door leading to the garden and windows too high for the raiders to look through, was a safer haven. To get there she had to pass the second window, where another burst caught her as she dragged a sheet from the bed to staunch the bleeding.

My mind could not contain the absolute horror of being alone, at the lowest ebb of night, and having automatic fire come at you from six feet away, showering glass into an echoing room lit only by muzzle flashes and burning curtains. Somehow Joan had found the strength to get to the bathroom and lie on the floor to hide. She had her mobile phone and called John Sutton, the security provider who lived in her guest house, just thirty yards away, but who that night was five hundred miles off in Dar es Salaam. Poor John had listened helplessly as she called to him for help. There were more shots, this time right through that steel outside bathroom door, and Joan's voice had grown weaker and then quiet. My heart pounded in my ears. 'Oh God, my Joan,' I called as I fought off a paralysing dread of going into the bathroom.

I dropped to my knees on the sticky, gore-covered floor, and moved the blood-drenched shroud slowly back from her face. She wore a strangely peaceful, resigned expression, as though she had accepted sleeping rough that night, as she had so many times before. She looked her age now, her hair had thinned and worry lines etched her face, but she was still a handsome, athletic woman who should have had many more years. Her glasses had slid down her nose. I smiled a heartbroken smile at the memory, pushed them back up for the last time, kissed her and finally broke down, the scalding tears coming up from a bottomless well of sorrow.

I buried Joan's ashes below a seedling fig tree in front of the house at Naivasha, with views looking over the lake she had loved so deeply, and scattered more at some of the beautiful places where we had camped, back in those first magical years we had shared. Joan had been elemental in my life, her energy and enthusiasm so vital for the success of all our ventures. I remembered the fearless way she had joined me in the water with hippos, or in the balloon over Kilimanjaro, the tireless way she had cared for the bongos and kept our camp running so I could concentrate on filming. In those years, filled with love, success and adventure, we had flown high and far together because, as I said to the great gathering at her memorial service, 'She was the wind beneath my wings.'

COUNTY LIBRARY SERVICE LOUTH

Epilogue

My life has come full circle. I am once again on the Serengeti – my spiritual home. I am gathering wildlife footage for a feature film about the Grzimeks' work tracking the great wildebeest migration, work that I had helped to film and that won an Oscar so long ago. From my camp, I look out on a land in many ways unchanged in the almost fifty years since I worked on that film as a young man. There are no signs of human activity, no roads, buildings or camps. They are out there, of course, many of them, but all hidden from this vantage point. I chose this site carefully, aware as I am of my need not to see those signs of 'progress', and how, over the years, it has become harder, or even impossible, to find such an unspoiled view.

My sunlit life has been spent in the most glorious countries, among the most spectacular wildlife, much of it in the company of Africans – wonderful, resilient people who are among the friendliest and most cheerful on earth. But that life has run parallel with a heartbreaking holocaust as wildlife conservation has proved to be a disastrous failure. The reasons are many, ranging from greed, myopia and failed policies to the exponential growth of the human population, which continues to sweep away wildlife and wild places. This is no place for a disquisition into that muddled and tragic history; suffice to say that the Serengeti is one place that has fulfilled its remit to protect the greatest wildlife show on earth and, looking out from my camp, I am happy to be back in the place where I feel so at home.

And now I have an even greater source of happiness. From my life with Joan and Jenny, the women with whom, in two totally different relationships, I spent most of my adult years, I learned that while I was not a very sociable animal, nor was I cut out to be a loner. I have always needed a mate, a companion, someone with whom to share my adventures. So the ultimate serendipity of my life was meeting Fran Michelmore, a biologist, artist, pilot and concert violinist – clever, vibrant and lovely. Born and schooled largely in England, Fran had worked with Gerald Durrell at his zoo in Jersey, studied mother-infant behaviour among captive gorillas and spent time in India studying tigers. She had come out to Africa in 1989 to build a continent-wide computer database on elephants, and since then had created a niche for herself conducting environmental studies in Kenya, Tanzania and Uganda. We had many friends in common but had only met a few times. All that changed when, one evening, watching her play her violin passionately at a performance of Handel's *Messiah* in Nairobi's cathedral, I realised I wanted nothing more than to spend whatever was left of my life with her. Fortunately she thought that wasn't an altogether bad idea.

So I went to Justin Hando, the African Warden of the Serengeti and a good friend, to ask if I could put up a camp in a remote area of his park and fly in a lot of friends to a home-made strip there. It was a pretty weighty request and when he asked why, I told him that the Serengeti was where my heart lived, and was where I wanted to get married. Having known both Joan and Jenny, Hando's warm response was typical of the African regard for anything involving family. '*Karibu!*' he said. 'Welcome! You belong here, so don't bother me with the details – just get on with it. Congratulations and *enda salama mzee*.' ('go in peace old man').

I knew of a giant fig tree, its trunk scratched up by lion and leopard, with branches spreading well over a hundred feet – the biggest in the Serengeti – which grew on top of a high plateau climbed by no roads, where only animals roamed. A safari operator friend put up a big comfortable camp below the plateau, and close by we made an airstrip by driving up and down to flatten the grass and make sure there were no holes or stumps. I then took some men up to the fig tree in the helicopter, and we hand-cleared another strip up there and cut the grass under that great tree. Our friends flew into the bottom strip in several planes, and after a night in camp we all flew up to the top strip, where Fran and I were married under that magical tree. Both airstrips were short and tricky, so we had to limit the champagne until we were all safely back in camp, but then the full moon came up, and the African camp staff came in singing and drumming with a special cake they had made for us. Ridiculously happy and surrounded by good friends in the land that I loved, all the pain in my life drained away and I felt I had finally emerged from a long, dark tunnel, back into the sunshine where I belonged.

After years of longing for children of my own, Fran has given me two wonderful sons, Myles and Rory, and after bottle-raising over forty species of wild animals I have finally been able to help raise a couple of human babies, delighting in every step of that long-delayed experience. Fran is doing a brilliant job home-schooling the boys in a dedicated tent, strung with horns, feathers, papier-mâché animals and finger paintings, and they are inquisitive, bright and bold, free to explore and grow among animals in a wild place with distant horizons and an element of danger – something I had feared my generation might be the last to experience.

Our camp is the most comfortable I have ever enjoyed, the helicopter is parked nearby, and the land that I know and love so well is spread out before us. In the faded blue distance are the Ngorongoro highlands and closer in those huge jumbled piles of granite, the Moru Kopjes. The plains are still green, but are drying out fast and the great herds are massing for their trek north. They pile up here every year, reluctant to enter the bush country after months on the open plains where it is easy to spot predators. Perhaps half a million animals, zebra and wildebeest, are spread over the green plain below us – like salt and pepper sprinkled on a pool table.

I watch, as alert as a lioness, as my cubs – two blond little boys – walk fearlessly from camp for a hundred yards out among the great rocks. They both lug their heavy backpacks stuffed with magnifying glasses, string, SAS survival booklets, various containers, knives, fossils and feathers – probably forbidden matches and, almost certainly, no water. One dangles binoculars, the other a jam jar, and waddling behind is their pet hyrax. They start turning over stones and peering into bushes, their golden heads coming together whenever they find something interesting to share. Their hornbill flies out to join them and tries to land on Rory's shoulder, but he is too engrossed and keeps shaking it off. Then comes the call that sends my heart soaring, 'Hey, Dad, quick, come and look at this!'

Myles has found a giant stick insect and it rests on his arm, its spindly legs reaching from wrist to elbow, its hooked claws dug into his flesh. The boys examine it with interest, reverence and several 'wows', observing that it has laid some eggs on its own rear legs, mistaking them for twigs – or maybe it was some other insect that made that mistake. Then they gently release it, taking as much pleasure from seeing it go as they did from observing it. I walk with them to check on the progress of two nightjar chicks that crouch, perfectly camouflaged, on the bare granite, and we search a narrow crack in the rock, hoping to find the tiger snake that once escaped us by sliding in there. I am so thankful to be seeing my world again through their eager eyes, and to feel the powerful reasons they give me to look for positive signs and to see that although the glass is certainly not half full nor is it yet completely empty.

I realise that, like many who have spent their lives fighting to save some remnant of our wildlife, I could so easily have accepted that invitation to bitterness. The threats are undiminished and the battle is as desperate as ever, but there are areas where the ground is being held and one of them is the Serengeti, that irreplaceable fragment of the great age of mammals that has survived to give us a glimpse of the Pleistocene. When Tanzania gained independence in 1961 the Serengeti was its only national park. Today, Tanzania has fifteen national parks or game reserves, several of them set up at the request of village councils, eager to have their own wildlife refuges. Some sixteen per cent of the country is set aside for conservation. That is admirable, but is not to say that all is well.

The corruption in both the hunting industry, and the authority meant to regulate it, enables poaching for meat and ivory that is widespread and out of control. The export of thousands of birds, tortoises, chameleons and small mammals is a squalid business with heavy losses between trapping and shipping, and is a huge drain on those populations. Despite management plans calling for a halt to construction inside the parks, important people have been allocated sites to build sprawling, totally inappropriate hotels and the authorities' urge to maximise income has allowed tented camps to spring up like mushrooms. A plan to build a highway cutting across the route of the wildebeest migration was only recently dropped in the face of international condemnation and offers fully to fund alternative, less damaging routes; but the plan has not gone away completely and could easily be resuscitated. Another project being pushed forward would build a soda extraction plant on Lake Natron, where most of Africa's flamingos breed. Either of these schemes, if implemented without full consideration of less disruptive alternatives, would negate much of the positive progress the country has made since independence.

Yet Tanzania, for all her many problems, has a better record than most African countries. The Grzimeks' film was entitled *Serengeti Shall Not Die*, and while the park may be afflicted by ailments such as poaching, encroachment and poorly regulated development, that title is still Tanzania's watchword and the Serengeti is valued as part of the nation's identity. Long may those sentiments last.

Driving with Fran and the boys in search of sequences for the film about the Grzimeks' work, I share with them the memories that come back in droves. I show them my very first campsite – the great acacia still spreading its shade – we climb among the roots of the fig tree where the mamba looked me over and peer down at the rock where the leopard bit my bum. Some days we all squeeze into a little hide and watch a huge crocodile stalking the drinking wildebeest, on others we take a picnic and walk along the Mbalangeti river, where the boys find fossils and make plaster casts of big-cat tracks in the sand. I don't know if there is such a thing as genetic memory, but I do believe that the human spirit feels at home on the savannah. The combination of short grass, so that predators can be spotted, and big, climbable trees that provide escape, is the sort of habitat that enabled our ancestors to venture out from the forests and begin the long march to humankind. In our city parks, with their wide lawns and spreading trees, we recreate microcosms of the landscape that those ancestors found safe and comforting – and that we still do.

With the boys we visit Olduvai Gorge and Laetoli, where Mary Leakey discovered the fossilised hominid footprints that I helped to explain decades ago – tracks that call to us over immense time, urging us to pause and look back. For all the world's crushing problems, we must never lose sight of the importance of places like the Serengeti. Here, on the most bountiful grasslands on earth, some two million animals ebb and flow with the seasons across a landscape little changed since our ancestors took those first uncertain steps out on to the savannah. We pattered into the Pleistocene and plodded through the next million years, but now man's boots are made for walking. Our footprints on the moon are one of mankind's most vivid icons, our tread has grown heavier and we no longer need fossilisation to give our trampling permanence. I believe the Serengeti, where we can see the beginnings of that long march, is one place where we should give it pause. To stand on those plains, dwarfed by the distances, surrounded by legions of animals trekking to the horizon, is to be transfixed by nature's abundance. To stand near those ancient footprints on the edge of the plain is to be overwhelmed by man's insignificance and uncertain future. Pressing close together for comfort, the tracks set out across our mighty continent, filled with danger and promise, and then – after thirty yards – they disappear over a cliff. Under that vault of uncaring sky it is a bleak allegory.

Perhaps, some day, a future generation that looks back and laughs at the ponderous silicon chip and the clumsy space shuttle will not need wilderness. May be we will outgrow our animal past and even our human present. But until that time, as mankind gropes for direction, I believe wild places like the Serengeti will be essential for our spiritual well-being. A place where we can rekindle primeval fires, look back at where our tracks have come from and perhaps work out where they ought to be going.

Acknowledgements

The actual writing of this book has been an almost solo operation, but the life it describes has been enriched by many individuals, and for their teachings, camaraderie, inspiration and love I am deeply grateful to all those I have mentioned in these pages. Thank you, too, to the many others unnamed: you are neither forgotten nor unappreciated. My special thanks to Wolfgang Weber for the fine sketches that grace many of these pages.

My grateful thanks to Jeremy Bradshaw, who was the book's midwife, persuading me I should and could write it, and supporting me through the whole process.

Thanks too to my agent Patrick Walsh, and editor at Chatto & Windus, Becky Hardie, who have patiently steered the often runaway vehicle of my writings back to the trail, and with whom several differences have been hammered out – some at white heat.

My wonderful family, Fran, Myles and Rory, have for a very long time good-naturedly put up with the grumpy old guy that an outdoors type becomes when fidgeting in front of a computer for hours every day, desperately trying to remember names and dates. Thank you for your love and understanding – and all the coffee.

Permissions

For permission to quote from John Masefield's 'Cargoes', thanks to the Society of Authors as the Literary Representative of the Estate of John Masefield, Ledbury, Herefordshire, HR8 1PN. 'The Cremation of Sam McGee' by Robert W. Service © The Estate of Robert W. Service.

All photographs are from the author's collection, excepting five images for which he is very grateful for permission to publish them: Sailing over Kilimanjaro © Alastair Grahame; orange lorry crossing a bridge in the Congo © Conrad Aveling; the new ways of crossing rivers in the Congo now that the bridges have gone © Thérèse Hart; Congo peacock © Norbert Rottcher; lions in the Ngorongoro Crater in front of the grave of Bernhard and Michael Grzimek © Christian Grzimek, Tierbild Okapia, Frankfurt; Rory, Fran, Myles and me © John Staley.

All illustrations are by Wolfgang Weber. The maps are by Jane Randfield and thanks also to Rob Shaw for setting the maps' place names.